Core Books in Advanced Mathematics

Vectors

Core Books in Advanced Mathematics

General Editor: C. PLUMPTON, Moderator in Mathematics,
University of London School Examinations Department;
formerly Reader in Engineering Mathematics,
Queen Mary College, University of London.

Advisory Editor: N. WARWICK

Titles available

Differentiation
Integration
Vectors
Curve sketching
Newton's Laws and Particle Motion
Mechanics of Groups of Particles
Methods of Trigonometry
Coordinate Geometry and Complex Numbers
Proof
Methods of Algebra
Statistics
Probability

Core Books in Advanced Mathematics

Vectors

Tony Bridgeman
Chief Examiner in Advanced Level Mathematics, University of London School Examinations Department, Senior Lecturer in Applied Mathematics, University of Liverpool.

P. C. Chatwin
Assessor in Advanced Level Mathematics, University of London School Examinations Department, Senior Lecturer in Applied Mathematics, University of Liverpool.

C. Plumpton
Moderator in Mathematics, University of London School Examinations Department; formerly Reader in Engineering Mathematics, Queen Mary College, University of London.

M
MACMILLAN

First published 1983
Reprinted 1984

Published by
MACMILLAN EDUCATION LTD
Houndmills, Basingstoke, Hampshire RG21 2XS
and London
Companies and representatives
throughout the world

Printed in Hong Kong

British Library Cataloguing in Publication Data
Bridgeman, Tony
Vectors. - (Core books in advanced mathematics)
1. Algebras, Linear 2. Vector spaces
I. Title II. Chatwin, P. C.
III. Plumpton, Charles IV. Series
512'.5 QA200

ISBN 0-333-31791-2

Contents

Preface

Advanced level mathematics syllabuses are once again undergoing changes of content and approach, following the revolution in the early 1960s which led to the unfortunate dichotomy between 'modern' and 'traditional' mathematics. The current trend in syllabuses for Advanced level mathematics now being developed and published by many GCE Boards is towards an integrated approach, taking the best of the topics and approaches of the modern and traditional, in an attempt to create a realistic examination target, through syllabuses which are maximal for examining and minimal for teaching. In addition, resulting from a number of initiatives, core syllabuses are being developed for Advanced level mathematics syllabuses, consisting of techniques of pure mathematics as taught in schools and colleges at this level.

The concept of a core can be used in several ways, one of which is mentioned above, namely the idea of a core syllabus to which options such as theoretical mechanics, further pure mathematics and statistics can be added. The books in this series are core books involving a different use of the core idea. They are books on a range of topics, each of which is central to the study of Advanced level mathematics; they form small core studies of their own, of topics which together cover the main areas of any single-subject mathematics syllabus at Advanced level.

Particularly at times when economic conditions make the problems of acquiring comprehensive textbooks giving complete syllabus coverage acute, schools and colleges and individual students can collect as many of the core books as they need, one or more, to supplement books already acquired, so that the most recent syllabus of, for example, the London, Cambridge, JMB and AEB GCE Boards, can be covered at minimum expense. Alternatively, of course, the whole set of core books gives complete syllabus coverage of single-subject Advanced level mathematics syllabuses.

The aim of each book is to develop a major topic of the single-subject syllabuses, giving essential book work and worked examples and exercises arising from the authors' vast experience of examining at this level and including actual past GCE questions also. Thus, as well as using the core books in either of the above ways, they would also be ideal for supplementing comprehensive textbooks in the sense of providing more examples and exercises, so necessary for preparation and revision for examinations on the Advanced level mathematics syllabuses offered by the GCE Boards.

Vectors are now regarded as an essential tool in mathematics, physics and engineering and the purpose of this book is to provide a systematic and coherent introduction to the subject of vector algebra. The many worked examples illustrate the elegance and practical convenience of vector methods in physical and geometrical problems. As with other branches of mathematics, vector algebra can only be mastered, and fully appreciated, by working conscientiously through the worked and unworked examples. The treatment assumes an elementary knowledge of calculus, a topic which is covered by companion volumes in this series.

The authors are grateful to the University of London Entrance and School Examinations Council (L) for permission to reproduce questions from past Advanced Level GCE papers.

Tony Bridgeman
P. C. Chatwin
C. Plumpton
Nov. 1982

1 An introduction to vectors

1.1 General background material

In applying mathematics, particularly in the fields of physics and engineering, we need to be able to associate numbers with quantities that are important.

We know from experience that many phenomena may be characterised by a single number. That is, provided the relevant units are correct, a single number x can represent

(i) the mass, or the volume, or the speed of a body,
(ii) the distance between two points,
(iii) the time taken for a man to run 100 m,

and many similar quantities. The representation is satisfactory because each of the quantities obeys the same rules of algebra as do real numbers. So, if we perform a given series of algebraic operations on a real number, we know that the result is equally applicable to any one of a very large set of physical quantities. Such quantities are called *scalars*.

On the other hand it is easy to think of quantities which are not completely characterised by a single number. For example, if we wish to move a body from a point A to a different point B, what has to be done depends not only on the distance we move the body but also on the direction in which we move it. Thus a displacement is completely specified by a scalar length *and* a direction. Similarly, the effect of a force on a body depends both on the magnitude of the force and the direction in which it is applied. Each of these quantities, displacement and force, always needs more than one number in order to describe it completely. As a further example, the velocity of a boat needs two numbers, such as a speed v and a bearing θ, whilst the velocity of an aeroplane needs three numbers, such as a speed v, a bearing θ and an elevation ϕ.

Now we consider whether it is possible to unify the mathematics of quantities which require more than one number for their specification. Fortunately we find that there is a large set of such quantities with similar specifications which behave in the same way when we perform operations like adding them together. These quantities are called *vectors*; but before defining a vector we first investigate the properties of linear displacements.

Consider the displacement of a point A into a point B along the straight line AB. This displacement may be represented diagrammatically by a directed straight line (see Fig. 1.1), and at this stage we denote it by $\overrightarrow{AB}$. Note however that the displacement is not necessarily associated with the point A; all it tells

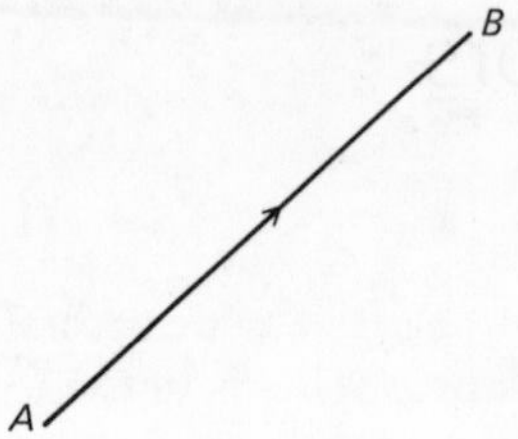

Fig. 1.1

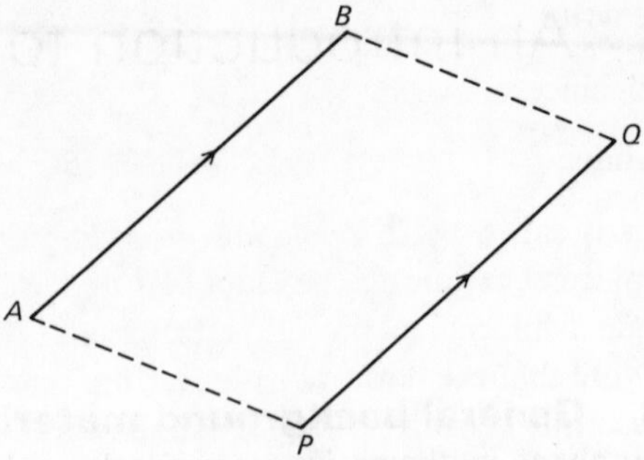

Fig. 1.2

us is the effect it has on A. The same displacement moves a point P into a point Q where the directed line segment $\overrightarrow{PQ}$ is equal in length and parallel to the directed line segment $\overrightarrow{AB}$ (see Fig. 1.2).

From Fig. 1.2 we can see that $APQB$ is a parallelogram. So the displacement is completely specified by a scalar magnitude and a direction. For such displacements we say that $\overrightarrow{AB}$ equals $\overrightarrow{PQ}$, and write

$$\overrightarrow{AB} = \overrightarrow{PQ}.$$

Simple algebraic operations on displacements are easily described. For instance, if we wish to displace A twice as far as B but in the same direction, it is natural to denote the corresponding displacement by $2\overrightarrow{AB}$ and to interpret this as changing the length (or magnitude) of the displacement by a factor of 2 without changing its direction. Similarly $\lambda\overrightarrow{AB}$, where λ is a positive scalar, may be interpreted as a displacement in the same direction as $\overrightarrow{AB}$ but with a magnitude λ times that of $\overrightarrow{AB}$. Such a multiple of a displacement is called a *scalar multiple*.

The operation of addition may similarly be defined by considering the result of two successive displacements. Given two displacements $\overrightarrow{AB}$ and $\overrightarrow{RS}$ we see from Fig. 1.3 that $\overrightarrow{AB}$ takes A to B and then if $\overrightarrow{BC} = \overrightarrow{RS}$ the second displacement takes B to C. So the net result of $\overrightarrow{AB}$ followed by $\overrightarrow{RS}$ is the displacement $\overrightarrow{AC}$.

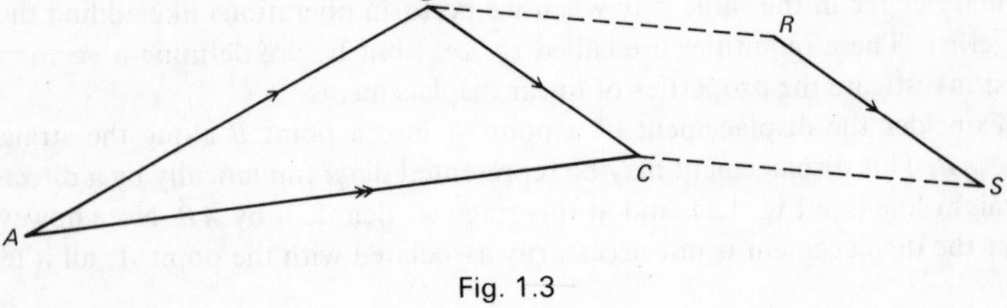

Fig. 1.3

We write $$\overrightarrow{AB} + \overrightarrow{RS} = \overrightarrow{AC},$$
and, since $$\overrightarrow{RS} = \overrightarrow{BC},$$
we have $$\overrightarrow{AB} + \overrightarrow{BC} = \overrightarrow{AC}. \tag{1.1}$$

This is called the *triangle law of addition* because the result of two successive displacements along the sides AB and BC of the triangle ABC is equal to a single displacement along the third side AC.

Note that the addition rule is often referred to also as the *parallelogram law of addition* because, if $ABCD$ are the vertices of a parallelogram taken in order, then

$$\overrightarrow{AB} + \overrightarrow{AD} = \overrightarrow{AC}.$$

However, from Fig. 1.4, $\overrightarrow{AD} = \overrightarrow{BC}$ and so we see that the parallelogram law and the triangle law are the same.

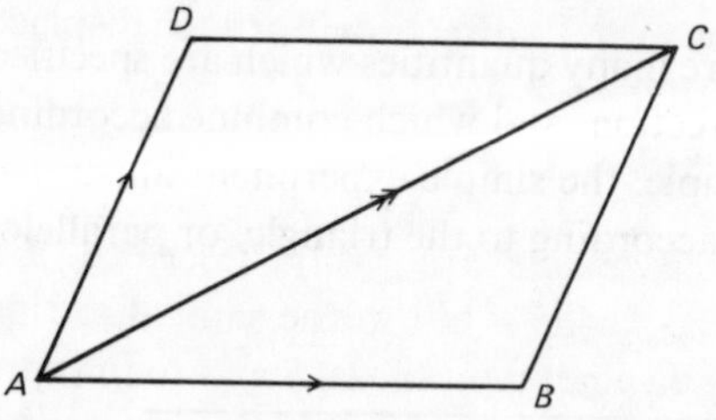

Fig. 1.4

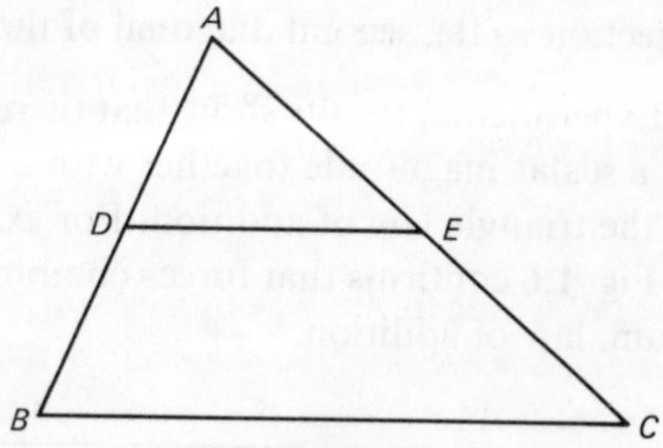

Fig. 1.5

Example 1 If A, B and C are any three distinct points and D, E are the mid-points of the line segments AB, AC respectively, show that $\overrightarrow{DE} = \frac{1}{2}\overrightarrow{BC}$. (Fig. 1.5)

Clearly, $$\overrightarrow{DE} = \overrightarrow{DA} + \overrightarrow{AE}.$$
But $$\overrightarrow{DA} = \tfrac{1}{2}\overrightarrow{BA} \quad \text{and} \quad \overrightarrow{AE} = \tfrac{1}{2}\overrightarrow{AC}$$
$$\Rightarrow \overrightarrow{DE} = \tfrac{1}{2}(\overrightarrow{BA} + \overrightarrow{AC}) = \tfrac{1}{2}\overrightarrow{BC}.$$

This proves the theorem that the line segment joining the mid-points of two sides of a triangle is parallel to the third side of the triangle and has half its length.

As the line AA has zero length the displacement $\overrightarrow{AA}$ leaves the point A undisplaced and so is naturally denoted by **0**. (The use of the bold notation is necessary to distinguish the zero displacement **0** from the number 0. See pages 5 and 6). Then, using equation (1.1),

$$\overrightarrow{AB} + \overrightarrow{BA} = \mathbf{0}.$$

Consequently, we write $\overrightarrow{BA} = -\overrightarrow{AB}$ and interpret $-\overrightarrow{AB}$ as a displacement with the same magnitude as $\overrightarrow{AB}$ but in the opposite direction to $\overrightarrow{AB}$. This enables an interpretation to be given to a scalar multiple of a displacement when the scalar is negative. For example, if $\lambda = -\mu$, where $\mu > 0$, the displacement $\lambda\overrightarrow{AB} = -\mu\overrightarrow{AB}$ is a displacement in the opposite direction to $\overrightarrow{AB}$ and with magnitude μ times that of $\overrightarrow{AB}$.

Example 2 Give geometrical descriptions of the displacements $\overrightarrow{AB} + \overrightarrow{AD}$ and $\overrightarrow{AB} - \overrightarrow{AD}$, where A, B, C, D are the vertices of a parallelogram.

By the parallelogram law of addition $\overrightarrow{AB} + \overrightarrow{AD}$ is equal to the displacement $\overrightarrow{AC}$, where AC is the diagonal of the parallelogram through A.

Now
$$\overrightarrow{AB} + \overrightarrow{BD} = \overrightarrow{AD}$$
$$\Rightarrow \overrightarrow{AB} - \overrightarrow{AD} = -\overrightarrow{BD} = \overrightarrow{DB}.$$

Hence $\overrightarrow{AB} - \overrightarrow{AD}$ is the displacement $\overrightarrow{DB}$, which is equal in magnitude and direction to the second diagonal of the parallelogram.

Experimental results show that there are many quantities which are specified by a scalar magnitude together with a direction, and which combine according to the triangle law of addition. For example, the simple experiment illustrated in Fig. 1.6 confirms that forces combine according to the triangle, or parallelogram, law of addition.

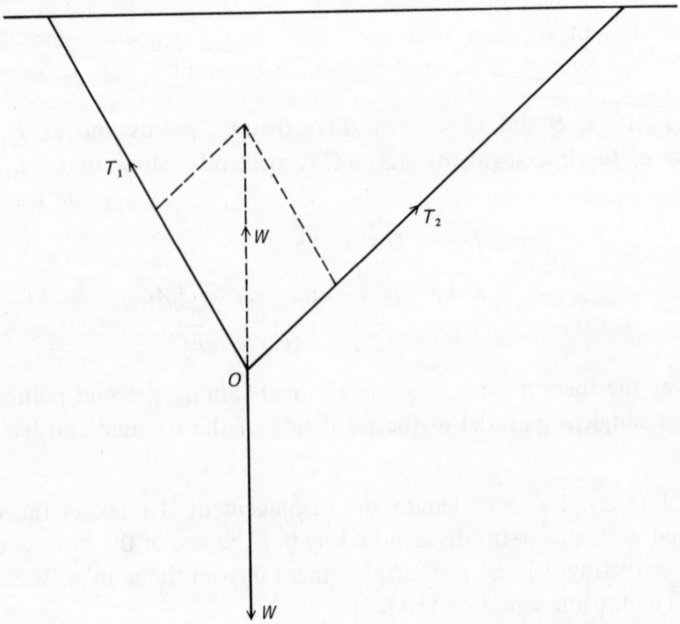

Fig. 1.6

In the figure, three strings are joined at a point O. One string carries a given weight W and the other two strings are held so as to support the weight. The supporting tensions T_1, T_2 act along the strings and their resultant must exactly balance the suspended weight. The tensions may be measured experimentally by introducing light spring balances. If now we draw lines, in the directions of the two supporting strings, whose lengths are equal in magnitude to the measured values of the tensions, the diagonal of the parallelogram determined by the two lines is found to be equal in magnitude and direction to the weight W. This illustrates that forces combine according to the parallelogram law of addition.

1.2 Definition of a vector

A *vector* is defined to be any quantity which is completely specified by a scalar magnitude and a direction, and whose sum with a second vector is obtained by the triangle law.

There are many such quantities but some of particular interest are displacement, velocity, acceleration, momentum, impulse, force and angular velocity. Any result involving vectors is equally applicable to each of these quantities and we shall see that the vector methods used are independent of whether the system under consideration is in one, two or three dimensions. Consequently the use of vectors removes any need to consider two- and three-dimensional problems separately.

Notation

We have already introduced the notation $\overrightarrow{AB}$ for a vector displacement but, whilst this may be retained, it is more convenient, and conventional, to use a single boldface letter **a**. In normal handwriting it is difficult to write a boldface letter and so it is essential that some other means be used to distinguish a vector quantity from a scalar quantity. This is usually done by simply underlining the letter, viz. $\underline{a}$ or $\underset{\sim}{a}$, whenever it is a vector quantity. In diagrams we continue to represent vectors by directed line segments, as for displacements, the length of the line being proportional to the magnitude of the vector and its direction being that of the vector.

The magnitude, or length, of the displacement $\overrightarrow{AB}$ is denoted by $|\overrightarrow{AB}|$, called 'mod $\overrightarrow{AB}$', and the magnitude of the vector **a** is similarly denoted by $|\mathbf{a}|$, or simply by a when there is no chance of confusion.

1.3 Algebraic operations on vectors

Equality of vectors

As an extension of the definition of equal displacements in §1.1, we say that two vectors **a** and **b** are equal if

(i) they have the same direction, and
(ii) they have the same magnitude, i.e. $|\mathbf{a}| = |\mathbf{b}|$.

In such a case we use the usual equality sign and write $\mathbf{a} = \mathbf{b}$.

Addition of vectors

As all vectors combine according to the triangle law of addition we may use diagrams to show that addition is

(i) commutative, that is $\mathbf{a} + \mathbf{b} = \mathbf{b} + \mathbf{a}$,

(ii) associative, that is $(\mathbf{a} + \mathbf{b}) + \mathbf{c} = \mathbf{a} + (\mathbf{b} + \mathbf{c})$.

The relevant diagrams are shown in Figs. 1.7 and 1.8. It should be noted that the displacements in Fig. 1.8 need not lie in one plane.

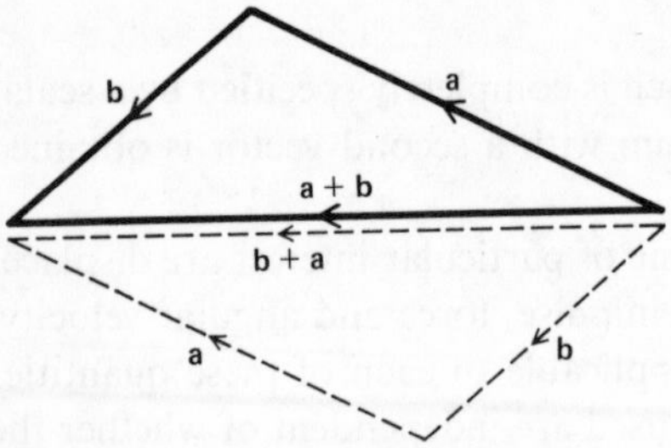

Fig. 1.7

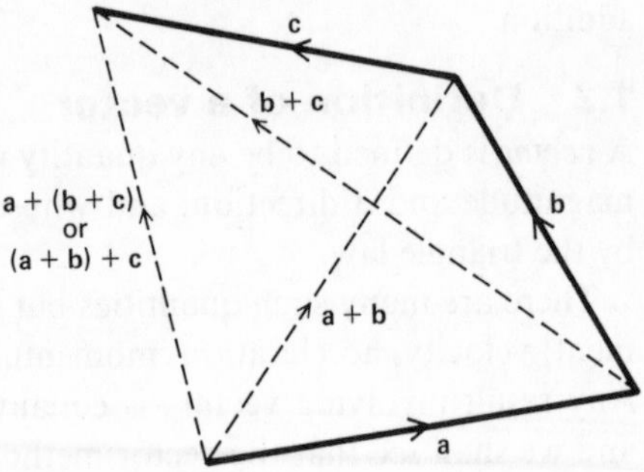

Fig. 1.8

Example 3 Given that A, B, C, D and E are any five distinct points in space, show that

$$\overrightarrow{AB} + \overrightarrow{BC} + \overrightarrow{CD} + \overrightarrow{DE} = \overrightarrow{AE}.$$

*By the triangle law of addition,

$$\overrightarrow{AB} + \overrightarrow{BC} = \overrightarrow{AC}$$

$$\Rightarrow \overrightarrow{AB} + \overrightarrow{BC} + \overrightarrow{CD} = (\overrightarrow{AB} + \overrightarrow{BC}) + \overrightarrow{CD} = \overrightarrow{AC} + \overrightarrow{CD} = \overrightarrow{AD}$$

$$\Rightarrow \overrightarrow{AB} + \overrightarrow{BC} + \overrightarrow{CD} + \overrightarrow{DE} = (\overrightarrow{AB} + \overrightarrow{BC} + \overrightarrow{CD}) + \overrightarrow{DE}$$

$$= \overrightarrow{AD} + \overrightarrow{DE} = \overrightarrow{AE}.$$

The null vector and vector subtraction

There is only one vector with zero magnitude; it is called the *null*, or *zero*, *vector*. It is written $\mathbf{0}$ and satisfies the equation

$$\mathbf{a} + \mathbf{0} = \mathbf{a},$$

for every vector $\mathbf{a}$.

---

* Where there is no diagram for a worked example, you are advised to draw one yourself so that the geometrical meaning is clear.

Then, given a vector **a**, we define the vector $(-\mathbf{a})$ by the equation

$$\mathbf{a} + (-\mathbf{a}) = \mathbf{0}.$$

So $(-\mathbf{a})$ is a vector equal in magnitude to **a** but directly opposite in direction. Hence, if $\mathbf{a} = \overrightarrow{AB}$, then $(-\mathbf{a}) = \overrightarrow{BA}$. Note that $(-\mathbf{a})$ may also be considered to be a scalar multiple of **a**, see p. 8.

We now introduce the concept of *vector subtraction*. Given two vectors **a** and **b** we define the difference $\mathbf{a} - \mathbf{b}$ by the equation

$$\mathbf{a} - \mathbf{b} = \mathbf{a} + (-\mathbf{b}).$$

To interpret this equation diagrammatically we let $\overrightarrow{OA}$, $\overrightarrow{OB}$ represent **a**, **b**, respectively, and let C be the fourth vertex of the parallelogram $OACB$ (see Fig. 1.9).

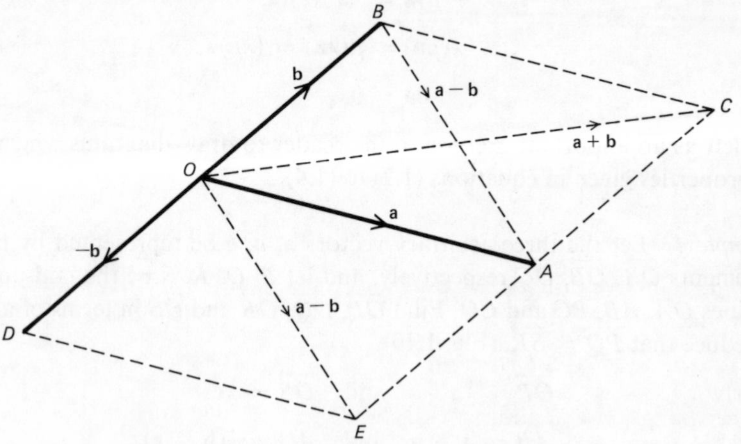

Fig. 1.9

Then $$\overrightarrow{OA} = \overrightarrow{OB} + \overrightarrow{BA}$$

and, adding $(-\overrightarrow{OB})$ to both sides,

$$\overrightarrow{OA} + (-\overrightarrow{OB}) = \overrightarrow{BA}$$

$$\Rightarrow \overrightarrow{BA} = \mathbf{a} + (-\mathbf{b}) = \mathbf{a} - \mathbf{b}.$$

Alternatively, let BO be produced to a point D where $OB = OD$, so that $\overrightarrow{OD} = (-\mathbf{b})$. Then, if E is the fourth vertex of the parallelogram $ODEA$, the parallelograms $OACB$ and $DEAO$ are congruent and $\overrightarrow{OE} = \overrightarrow{BA}$.

However, $$\overrightarrow{OE} = \overrightarrow{OA} + \overrightarrow{OD} = \mathbf{a} + (-\mathbf{b}) = \mathbf{a} - \mathbf{b}$$

$$\Rightarrow \overrightarrow{BA} = \mathbf{a} - \mathbf{b}.$$

Consequently, if **a** and **b** are represented by two adjacent sides of a parallelogram, the two diagonals of the parallelogram represent **a** + **b** and **a** − **b** (see example 2, p. 4).

Scalar multiple of a vector

The *scalar multiple* of a displacement may be generalised to include all vectors. That is, given a positive scalar λ and a vector **a**, the vector $\lambda\mathbf{a}$ is defined to be a vector of magnitude $\lambda|\mathbf{a}|$ and direction the same as that of **a**. Similarly, if $\lambda = -\mu$ is a negative scalar, so $\mu > 0$, the vector $\lambda\mathbf{a} = -\mu\mathbf{a}$ has magnitude $-\lambda|\mathbf{a}| = \mu|\mathbf{a}|$ and direction opposite to that of **a**. This clearly provides a definition of $-\mathbf{a}$ consistent with that given above.

When λ, μ are arbitrary scalars and **a**, **b** are arbitrary vectors, it is easily shown that scalar multiplication has the following properties.

$$\lambda(\mathbf{a} + \mathbf{b}) = \lambda\mathbf{a} + \lambda\mathbf{b}, \tag{1.2}$$

$$(\lambda + \mu)\mathbf{a} = \lambda\mathbf{a} + \mu\mathbf{a}, \tag{1.3}$$

$$\lambda(\mu\mathbf{a}) = \mu(\lambda\mathbf{a}) = (\lambda\mu)\mathbf{a}, \tag{1.4}$$

$$0\mathbf{a} = \mathbf{0}. \tag{1.5}$$

It is left as an important exercise to the reader to draw diagrams which verify the properties given in equations (1.2) to (1.4).

Example 4 Let the three arbitrary vectors **a**, **b**, **c** be represented by the displacements $\overrightarrow{OA}$, $\overrightarrow{OB}$, $\overrightarrow{OC}$ respectively, and let P, Q, R, S be the mid-points of the lines OA, AB, BC and CO. Find $\overrightarrow{OP}$, $\overrightarrow{OQ}$, $\overrightarrow{OR}$ and $\overrightarrow{OS}$ in terms of **a**, **b** and **c**. Deduce that $\overrightarrow{PQ} = \overrightarrow{SR}$. (Fig. 1.10)

Clearly $\qquad \overrightarrow{OP} = \tfrac{1}{2}\mathbf{a} \quad$ and $\quad \overrightarrow{OS} = \tfrac{1}{2}\mathbf{c}.$

Now $\qquad \overrightarrow{AB} = \mathbf{b} - \mathbf{a} \quad$ and $\quad \overrightarrow{AQ} = \tfrac{1}{2}(\mathbf{b} - \mathbf{a}).$

$$\Rightarrow \overrightarrow{OQ} = \overrightarrow{OA} + \overrightarrow{AQ} = \mathbf{a} + \tfrac{1}{2}(\mathbf{b} - \mathbf{a}) = \tfrac{1}{2}(\mathbf{a} + \mathbf{b}).$$

Similarly we see that $\overrightarrow{OR} = \tfrac{1}{2}(\mathbf{b} + \mathbf{c})$.

But $\qquad \overrightarrow{PQ} = \overrightarrow{OQ} - \overrightarrow{OP}$

$$\Rightarrow \overrightarrow{PQ} = \tfrac{1}{2}(\mathbf{a} + \mathbf{b}) - \tfrac{1}{2}\mathbf{a} = \tfrac{1}{2}\mathbf{b}.$$

Similarly $\qquad \overrightarrow{SR} = \overrightarrow{OR} - \overrightarrow{OS} = \tfrac{1}{2}(\mathbf{b} + \mathbf{c}) - \tfrac{1}{2}\mathbf{c} = \tfrac{1}{2}\mathbf{b}.$

Thus $\overrightarrow{PQ} = \overrightarrow{SR}$, and this means that the line PQ is equal and parallel to the line SR, so that $PQRS$ is a parallelogram. This proves the general result that the mid-points of the sides of *any* quadrilateral (even if it is skew, that is if there is no plane containing all four vertices) form the vertices of a parallelogram.

Example 5 Given that Q, R, T are the mid-points of the sides AB, BC, CA respectively of a triangle, show that $\overrightarrow{AR} + \overrightarrow{BT} + \overrightarrow{CQ} = \mathbf{0}$.

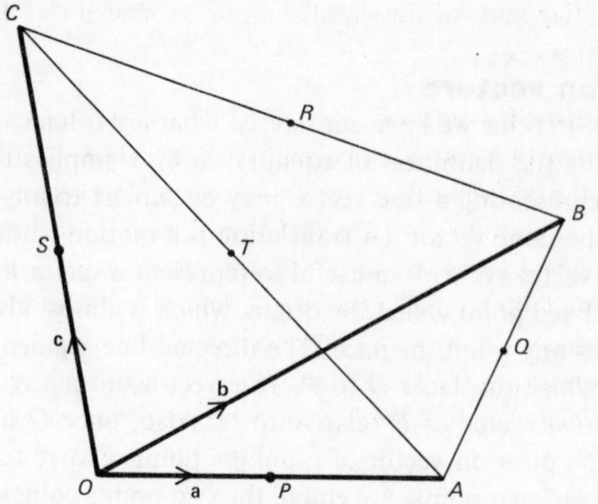

Fig. 1.10

Using the same notation and diagram as in Example 4, we see that

$$\overrightarrow{OT} = \overrightarrow{OA} + \tfrac{1}{2}\overrightarrow{AC} = \mathbf{a} + \tfrac{1}{2}(\mathbf{c} - \mathbf{a}) = \tfrac{1}{2}(\mathbf{a} + \mathbf{c}),$$

with similar results for $\overrightarrow{OQ}$ and $\overrightarrow{OR}$. It follows that

$$\overrightarrow{AR} = \overrightarrow{OR} - \overrightarrow{OA} = \tfrac{1}{2}(\mathbf{b} + \mathbf{c}) - \mathbf{a},$$
$$\overrightarrow{BT} = \overrightarrow{OT} - \overrightarrow{OB} = \tfrac{1}{2}(\mathbf{a} + \mathbf{c}) - \mathbf{b},$$
$$\overrightarrow{CQ} = \overrightarrow{OQ} - \overrightarrow{OC} = \tfrac{1}{2}(\mathbf{a} + \mathbf{b}) - \mathbf{c}.$$

Substituting the above values,

$$\overrightarrow{AR} + \overrightarrow{BT} + \overrightarrow{CQ} = \mathbf{0}.$$

The angle between two vectors

Given any two non-zero vectors **a** and **b** we may select any point O and represent **a** and **b** by the directed line segments $\overrightarrow{OA}$ and $\overrightarrow{OB}$ respectively. Then the *angle* between **a** and **b** is $\theta = A\hat{O}B$, where $0 \leqslant \theta \leqslant \pi$, assuming that θ is measured in radians. Consequently, if $\theta = 0$ the vectors are parallel and if $\theta = \pi/2$ they are perpendicular. If either, or both, **a** and **b** are **0**, the angle between them is not defined, or ever needed.

Example 6 Show that if **a** and **b** are unit vectors, that is $|\mathbf{a}| = |\mathbf{b}| = 1$, then $\mathbf{a} + \mathbf{b}$ and $\mathbf{a} - \mathbf{b}$ are perpendicular.

Let $\overrightarrow{OA}$, $\overrightarrow{OB}$ represent the vectors **a** and **b**. Then as **a** and **b** are unit vectors, $|\mathbf{a}| = |\mathbf{b}| = 1$ and so $|\overrightarrow{OA}| = |\overrightarrow{OB}| = 1$. Consequently $\overrightarrow{OA}$ and $\overrightarrow{OB}$ form adjacent sides of a rhombus. The diagonals of a rhombus are perpendicular and so, by example 2 (p. 4), $\mathbf{a} + \mathbf{b}$ and $\mathbf{a} - \mathbf{b}$ are perpendicular.

1.4 Position vectors

In the previous sections we have considered what are often called *free vectors*. For such vectors the definition of equality, in §1.3, implies that the directed line segment representing a free vector may be subject to any translation and still represent the same vector. (A translation is a motion without rotation.)

When using vectors it is often useful to represent a vector by a line segment beginning at a fixed point called the origin, which is almost always denoted by O. Suppose P is any point in space. The directed line segment $\overrightarrow{OP}$ represents a vector, that which displaces O to P. This vector, normally denoted by $\mathbf{r}$, is called the *position vector* of P relative to O. Also, once O has been chosen, any vector is the position vector of a unique point relative to O. Thus, if the position vectors of two points are equal, the two points coincide.

Position vectors are used in mechanics to describe the motion of a particle which is subject to a set of (vector) forces. In geometry they enable many shapes to be represented by simple equations and often help to provide elegant proofs of theorems. For example, if $\mathbf{r}$ is the position vector of an arbitrary point P with respect to O and if d is a positive constant, the equation $|\mathbf{r}| = d$ tells us that P is a distance d from O. So P can be any point on the surface of the sphere with centre O and radius d. Hence $|\mathbf{r}| = d$ is said to be a vector equation of the sphere.

Example 7 Assume that point masses $m_1, m_2, \ldots, m_k$ are located at the points $A_1, A_2, \ldots, A_k$ whose position vectors relative to an origin O are $\mathbf{a}_1, \mathbf{a}_2, \ldots, \mathbf{a}_k$. The *centroid* (often called the *centre of mass*) of the masses is the name for the point G whose position vector, relative to O, is given by $\mathbf{g}$, where

$$\mathbf{g} = \frac{m_1\mathbf{a}_1 + m_2\mathbf{a}_2 + \ldots + m_k\mathbf{a}_k}{m_1 + m_2 + \ldots + m_k}.$$

We show now that the position of the point G is independent of the choice of origin O.

Let O' be a point whose position vector relative to O is $\mathbf{h}$. Then taking O' as a new origin, the position vectors of $A_1, A_2, \ldots, A_k$ are $\mathbf{a}_1 - \mathbf{h}, \mathbf{a}_2 - \mathbf{h}, \ldots, \mathbf{a}_k - \mathbf{h}$, respectively. The centroid is then the point G', where

$$\overrightarrow{O'G'} = \frac{m_1(\mathbf{a}_1 - \mathbf{h}) + m_2(\mathbf{a}_2 - \mathbf{h}) + \ldots + m_k(\mathbf{a}_k - \mathbf{h})}{m_1 + m_2 + \ldots + m_k}$$

$$= \frac{m_1\mathbf{a}_1 + m_2\mathbf{a}_2 + \ldots + m_k\mathbf{a}_k}{m_1 + m_2 + \ldots + m_k} - \mathbf{h}$$

$$\Rightarrow \overrightarrow{O'G'} = \mathbf{g} - \mathbf{h} = \overrightarrow{O'G}.$$

Hence the points G and G' coincide, as they both have the same position vector relative to O', and we deduce that the position of the centroid is independent of the origin.

Exercise 1

1 *ABCD* is a parallelogram with $\overrightarrow{AB} = \mathbf{a}$ and $\overrightarrow{AD} = \mathbf{b}$. The point *E* is such that $\overrightarrow{DE} = 2\mathbf{b}$. Draw a sketch to illustrate this, and express the vectors $\overrightarrow{AE}$, $\overrightarrow{AC}$ and $\overrightarrow{EC}$ in terms of **a** and **b**. (L)

2 In the quadrilateral *OABC*, *D* is the mid-point of *BC* and *G* is the point on *AD* such that $AG:GD = 2:1$. Given that $\overrightarrow{OA} = \mathbf{a}$, $\overrightarrow{OB} = \mathbf{b}$ and $\overrightarrow{OC} = \mathbf{c}$, express $\overrightarrow{OD}$ and $\overrightarrow{OG}$ in terms of **a**, **b**, and **c**. (L)

3 The vertices *X*, *Y*, *Z* of a triangle have position vectors **x**, **y**, **z** respectively. *S* is the mid-point of *YZ* and *G* is the point dividing *XS* internally in the ratio 2 : 1. Write down the position vectors of *S* and *G*. Deduce that the three medians of a triangle are concurrent. (L)

4 The points *X*, *Y*, *Z* are the mid-points of the sides *QR*, *RP* and *PQ* respectively of triangle *PQR* and $\overrightarrow{QR} = \mathbf{p}$, $\overrightarrow{RP} = \mathbf{q}$. Find the vectors which represent the sides of the triangle *XYZ*. Find also vectors which represent $\overrightarrow{RX}$, $\overrightarrow{RY}$, $\overrightarrow{RZ}$ and $\overrightarrow{RG}$, where *G* is the centroid of the triangle *PQR*. (L)

5 The point *P* is on the side *AB* of a triangle *OAB* such that $3\overrightarrow{AP} = 2\overrightarrow{PB}$. The position vectors of *A* and *B* are **a** and **b** respectively relative to the origin, *O*. Express $\overrightarrow{OP}$ in terms of **a** and **b**.

The medians of the triangle *OAB* meet at *G*. Express $\overrightarrow{OG}$ in terms of **a** and **b**.

Given that *C* is the point on *OB* such that $OC = \frac{1}{4}OB$, show that *P*, *G* and *C* lie in the same straight line. (L)

6 In *OAB*, $\overrightarrow{OA} = 6\mathbf{a}$ and $\overrightarrow{OB} = 6\mathbf{b}$. The mid-point of *OA* is *M* and the point *P* lies in *AB* such that $AP:PB = 2:1$. The mid-point of *OP* is *N*.
(a) Calculate, in terms of **a** and **b**, the vectors $\overrightarrow{AB}$, $\overrightarrow{OP}$ and $\overrightarrow{MN}$.
(b) Show that the area of the quadrilateral *AMNP* is half the area of *OAB*.
(c) The line *AN* produced meets *OB* at *C*. Given that $\overrightarrow{OC} = k\mathbf{b}$, find the value of *k*. (L)

7 (i) The position vectors with respect to the origin *O* of three points *A*, *D* and *B* are respectively **a**, **d** and **b**. The points *A*, *D* and *B* lie on a straight line so that $4\mathbf{a} + 3\mathbf{b} = 7\mathbf{d}$. Find the ratio of the length of *AD* to the length of *DB*.
(ii) *R* is the mid-point of a line *PQ* and **p** and **q** are the position vectors of *P* and *Q* with respect to the origin *O*. *PE* and *QD* are medians of the triangle *OPQ*. Find $\overrightarrow{OR}$, $\overrightarrow{PE}$, $\overrightarrow{QD}$ in terms of **p** and **q**. (L)

8 (i) Three points *P*, *Q* and *R* have position vectors **p**, **q** and $k(2\mathbf{p} + \mathbf{q})$ respectively, relative to a fixed origin. Find the numerical value of *k* if
(a) $\overrightarrow{QR}$ is parallel to **p**,
(b) $\overrightarrow{PR}$ is parallel to **q**,
(c) *P*, *Q* and *R* are collinear.
(ii) A hexagon *OABCDE* has its pairs of opposite sides parallel and equal. The position vectors of *A*, *B* and *C* relative to *O* are **a**, $\mathbf{a} + \mathbf{b}$, $\mathbf{a} + \mathbf{b} + \mathbf{c}$ respectively. Find the position vectors of *D* and *E* relative to *O*. (L)

9 In $\triangle AOC$, *P* and *Q* are the mid-points of *OA* and *OC* respectively. The position vectors of *A* and *C* referred to the origin *O* are **a** and **c** respectively.
(a) Show that $PQ = \frac{1}{2}(\mathbf{c} - \mathbf{a})$.
(b) If **x** is the position vector of the point *G* where the medians of $\triangle AOC$ meet, show that $\mathbf{x} = \frac{1}{3}(\mathbf{a} + \mathbf{c})$.
(c) If *D* is any other point with position vector **d** and *R* is the mid-point of *CD*, show that the position vector of *S*, the mid-point of *PR*, is

$$\mathbf{s} = \frac{\mathbf{a} + \mathbf{c} + \mathbf{d}}{4}.$$

(d) Show that S lies on GD and $3GS = SD$. (L)

10 Find the angle between the vectors **a** and **b** given that

$$|\mathbf{a}| = 3, \quad |\mathbf{b}| = 5, \quad |\mathbf{a} - \mathbf{b}| = 7.$$ (L)

11 The vectors **a** and **b** are such that $|\mathbf{a}| = 3$ and $|\mathbf{b}| = 5$. Calculate the magnitude of $\mathbf{a} + \mathbf{b}$ given that the angle between **a** and **b** is $\pi/3$. (L)

12 Two sides $\overrightarrow{AB}$ and $\overrightarrow{BC}$ of a regular hexagon $ABCDEF$ are represented by the vectors **p** and **q**. Show that $\overrightarrow{CD}$ is represented by $\mathbf{q} - \mathbf{p}$, and write down, in terms of **p** and **q**, vectors which represent $\overrightarrow{DE}$, $\overrightarrow{BD}$, $\overrightarrow{BF}$.

13 Two points A and B have position vectors **a** and **b** respectively. A point P divides AB internally in the ratio $\lambda : \mu$. Show that it has position vector

$$\frac{\mu\mathbf{a} + \lambda\mathbf{b}}{\lambda + \mu}.$$

Three non-collinear points A, B and C have position vectors **a**, **b** and **c** respectively. The point X divides AB internally in the ratio 2 : 1. The point Y divides BC internally in the ratio 2 : 1. Show that XY produced meets AC produced at the point with position vector $\frac{1}{3}(4\mathbf{c} - \mathbf{a})$.

14 The points A and B have position vectors **a** and **b** respectively relative to an origin O, which does not lie on AB. Three points X, Y, Z have position vectors $\frac{1}{4}\mathbf{a} + \frac{3}{4}\mathbf{b}$, $2\mathbf{a} - \mathbf{b}$, $\mathbf{a} + 3\mathbf{b}$ respectively. Show, in one diagram, the positions of A, B, X, Y and Z.

15 In the tetrahedron $OABC$, the mid-point of BC is X. The point G on AX is such that $AG : GX = 2 : 1$. Given that $\overrightarrow{OA} = \mathbf{a}$, $\overrightarrow{OB} = \mathbf{b}$ and $\overrightarrow{OC} = \mathbf{c}$, express $\overrightarrow{OG}$ in terms of **a**, **b** and **c**.

16 Given a quadrilateral $OABC$ (not necessarily plane), in which P, Q, R and S are the mid-points of OA, AB, BC and CO respectively and **a**, **b**, **c** are the position vectors of the points A, B, C referred to O,

(a) find in terms of **a**, **b** and **c** the position vectors of P, Q, R and S;

(b) find $\overrightarrow{PQ}$ and $\overrightarrow{SR}$ in terms of **b**.

(c) State the conclusions which can be drawn about

(i) the line joining the mid-points of two sides of a triangle,

(ii) the quadrilateral formed by joining the mid-points of the four sides of a quadrilateral.

17 The vertices A, B and C of a triangle have position vectors **a**, **b** and **c** respectively. The point P on BC is such that $BP : PC = 3 : 1$. The point Q on CA is such that $CQ : QA = 2 : 3$. The point R on BA produced is such that $BR : AR = 2 : 1$. The position vectors of P, Q and R are **p**, **q** and **r** respectively. Express **q** in terms of **p** and **r**, and hence, or otherwise, show that P, Q and R are collinear. State the value of PQ/QR.

18 A, B, C are three non-collinear points with position vectors **a**, **b**, **c** respectively. Given that P, Q, R are points on BC, CA and AB respectively such that $BP : PC = CQ : QA = AR : RB = 1 : 2$, show that the point X with position vector $\frac{1}{7}(2\mathbf{a} + \mathbf{b} + 4\mathbf{c})$ lies on both BQ and CR. Write down the position vectors of the other vertices of the triangle XYZ formed by the lines AP, BQ, CR, and deduce that the point of concurrence of the medians of this triangle is G, the centroid of the triangle ABC.

19 The points A, B, C have position vectors **a**, **b**, **c**. Points X, Y, Z divide BC, CA and

AB internally in the ratio $s:1$, and the lines *AX*, *BY*, *CZ* intersect in pairs in the points *P*, *Q*, *R*. By first showing that the position vector of *X* is $(\mathbf{b} + s\mathbf{c})/(1 + s)$, find the position vectors of *P*, *Q* and *R*.

20 If two vectors $\mathbf{p}$ and $\mathbf{q}$ are equal in magnitude, prove that the vectors $\mathbf{p} + \mathbf{q}$ and $\mathbf{p} - \mathbf{q}$ are perpendicular.

The vertices *P*, *Q*, *R* of a triangle are defined by the position vectors $\mathbf{p}$, $\mathbf{q}$, $\mathbf{r}$ respectively, referred to the circumcentre *O* as the origin. Deduce from the above result that the point with position vector $\mathbf{p} + \mathbf{q} + \mathbf{r}$ is the orthocentre *H* of the triangle. (The orthocentre is the point of intersection of the perpendiculars from the vertices to the opposite sides of the triangle.)

2 Dimensions and bases

2.1 Displacements in a plane

Again we consider displacements as a particular example of vectors and, in the first instance, restrict ourselves to displacements in a fixed plane.

Let $\mathbf{a} = \overrightarrow{PQ}$ and $\mathbf{b} = \overrightarrow{RS}$ be two given non-zero displacements in this plane. If $\overrightarrow{PQ}$ and $\overrightarrow{RS}$ are parallel, then $\mathbf{a}$ is a multiple of $\mathbf{b}$, so that there is a scalar λ such that $\mathbf{a} = \lambda\mathbf{b}$. In this case the two coplanar displacements are said to be *linearly dependent*. On the other hand, when $\overrightarrow{PQ}$ and $\overrightarrow{RS}$ are not parallel, $\mathbf{a}$ is not a scalar multiple of $\mathbf{b}$, and the displacements are said to be *linearly independent*.

Assume that $\mathbf{a}$ and $\mathbf{b}$ are linearly independent coplanar displacements, and let $\mathbf{c} = \overrightarrow{AB}$ be any other displacement in the plane. Now draw two lines in the plane, one through A and parallel to $\overrightarrow{PQ}$, and the second through B and parallel to $\overrightarrow{RS}$. The two lines will intersect in a point C (see Fig. 2.1). As, by construction, $\overrightarrow{AC}$ is parallel to $\mathbf{a}$ and $\overrightarrow{BC}$ is parallel to $\mathbf{b}$, there are scalars λ, μ such that $\overrightarrow{AC} = \lambda\mathbf{a}$ and $\overrightarrow{CB} = \mu\mathbf{b}$. From Fig. 2.1, and the triangle law of addition,

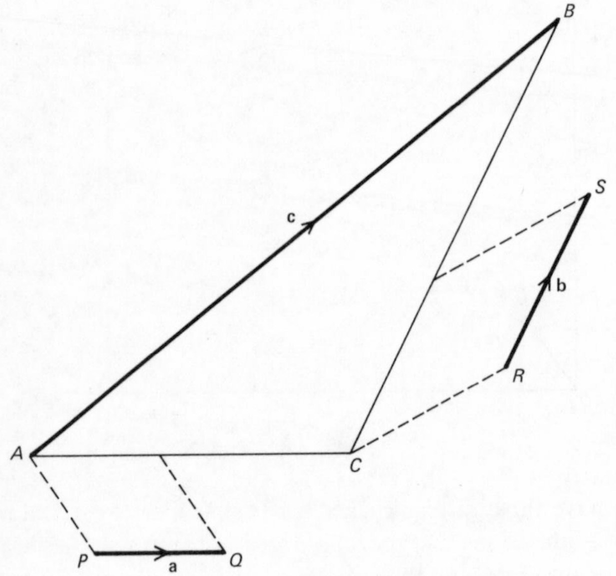

Fig. 2.1

$$\overrightarrow{AB} = \overrightarrow{AC} + \overrightarrow{CB}$$

or,

$$\mathbf{c} = \lambda\mathbf{a} + \mu\mathbf{b}.$$

So we see that every displacement in the plane is equal to the sum of a displacement proportional to **a** and a displacement proportional to **b**. That is, each displacement is a linear combination of the two linearly independent displacements **a** and **b**. Because each displacement can be so expressed in terms of two linearly independent displacements, the plane is said to have *dimension* two. The displacements **a** and **b** are said to form a *basis* for the set of coplanar displacements and the scalars λ, μ are called the *components* of the displacement **c** relative to the basis **a** and **b**.

In a plane, it is convenient to introduce Cartesian axes Oxy, where O is the origin relative to which position vectors are defined, and to adopt a basis which consists of a displacement **i** of unit length along the x-axis and a displacement **j** of unit length along the y-axis (see Fig. 2.2). It follows that, if **r** denotes the position vector of a point $R(x, y)$ in the plane, then $\mathbf{r} = x\mathbf{i} + y\mathbf{j}$. Similarly, if the displacement $\mathbf{a} = \overrightarrow{PQ}$, where P and Q have coordinates (x_1, y_1) and (x_2, y_2) respectively, then

$$\mathbf{a} = (x_2 - x_1)\mathbf{i} + (y_2 - y_1)\mathbf{j}.$$

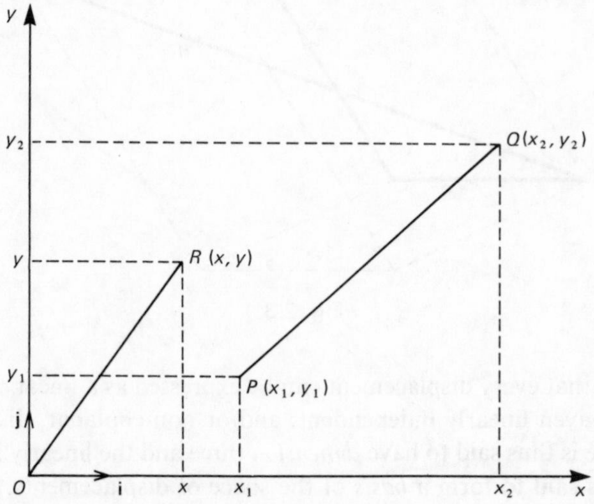

Fig. 2.2

When we have three displacements **a**, **b**, **c** and there exist scalars λ, μ such that $\mathbf{c} = \lambda\mathbf{a} + \mu\mathbf{b}$, we say that the set of three displacements $\{\mathbf{a}, \mathbf{b}, \mathbf{c}\}$ is *linearly dependent*, because **c** depends linearly on **a** and **b**. (Note that linear dependence is a property of the set; for, if $\lambda \neq 0$, we have $\mathbf{a} = \lambda'\mathbf{c} + \mu'\mathbf{b}$, where $\lambda' = \lambda^{-1}$

and $\mu' = -\mu\lambda^{-1}$.) We deduce that any set of three coplanar displacements is always linearly dependent.

Similarly, if there are no scalars such that $\mathbf{c} = \lambda\mathbf{a} + \mu\mathbf{b}$, we say that the set $\{\mathbf{a}, \mathbf{b}, \mathbf{c}\}$ is *linearly independent*. Clearly this means that the displacements are not coplanar.

2.2 Displacements in space

We now remove the restriction that the displacements must lie in a plane and consider displacements in any direction in real (i.e. ordinary) space.

Let **a**, **b**, **c** be three linearly independent displacements, which must be non-coplanar, and let **d** be any other displacement. Using Fig. 2.3 we see that it is possible to construct a skew quadrilateral whose sides represent the displacements $\lambda\mathbf{a}$, $\mu\mathbf{b}$, $\nu\mathbf{c}$ and **d**, where λ, μ, ν are scalars. In fact, the quadrilateral forms the diagonal and three non-parallel edges of a parallelepiped generated by three concurrent edges which represent $\lambda\mathbf{a}$, $\mu\mathbf{b}$ and $\nu\mathbf{c}$. Successive applications of the triangle law of addition lead to the result

$$\mathbf{d} = \lambda\mathbf{a} + \mu\mathbf{b} + \nu\mathbf{c}. \tag{2.1}$$

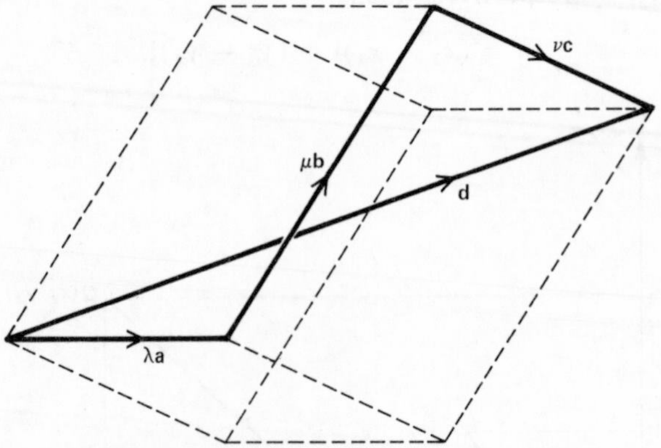

Fig. 2.3

We conclude that every displacement can be expressed as a linear combination of the three given linearly independent, and/or non-coplanar, displacements. Normal space is thus said to have *dimension* three and the linearly independent set $\{\mathbf{a}, \mathbf{b}, \mathbf{c}\}$ is said to form a *basis* of the space of displacements. Also λ, μ, ν are called the *components* of **d** relative to the basis $\{\mathbf{a}, \mathbf{b}, \mathbf{c}\}$.

Example 1 Prove that, for a given basis, the components of a vector are unique.

With the basis $\{\mathbf{a}, \mathbf{b}, \mathbf{c}\}$, let a vector **d** be given by equation (2.1). Then, if **d** has alternative components λ', μ', ν',

$$\mathbf{d} = \lambda'\mathbf{a} + \mu'\mathbf{b} + \nu'\mathbf{c}.$$

Subtracting (2.1),

$$(\lambda - \lambda')\mathbf{a} + (\mu - \mu')\mathbf{b} + (\nu - \nu')\mathbf{c} = \mathbf{0}$$

$$\Rightarrow (\lambda' - \lambda)\mathbf{a} = (\mu - \mu')\mathbf{b} + (\nu - \nu')\mathbf{c}$$

and, if $\lambda \neq \lambda'$,

$$\mathbf{a} = \mu''\mathbf{b} + \nu''\mathbf{c},$$

where $\mu'' = (\mu - \mu')/(\lambda' - \lambda)$ and $\nu'' = (\nu - \nu')/(\lambda' - \lambda)$. This contradicts the fact that **a**, **b**, **c** are linearly independent, and we deduce that $\lambda = \lambda'$. Similarly $\mu = \mu'$, $\nu = \nu'$ and so the components of **d** are unique.

In three-dimensional space we normally introduce cartesian axes $Oxyz$ and adopt a basis **i**, **j**, **k** which consists of displacements of unit length along the x, y and z-axes respectively. The corresponding components are naturally called cartesian components. It follows that, if **r** is the position vector of a point $R(x, y, z)$, the cartesian components of **r** are x, y, z and

$$\mathbf{r} = x\mathbf{i} + y\mathbf{j} + z\mathbf{k}.$$

2.3 Vectors in space

The arguments of §2.1 and §2.2 are applicable not just to displacements but to vectors generally. Thus, any vector **a** may be expressed in the form

$$\mathbf{a} = a_1\mathbf{i} + a_2\mathbf{j} + a_3\mathbf{k}, \tag{2.2}$$

where **i**, **j**, **k** are unit vectors along the coordinate axes and a_1, a_2, a_3 are scalars. The vector **a** could similarly be expressed as a linear combination of any three linearly independent vectors. However, it is usually more convenient to use **i**, **j**, **k** as the basis vectors. Such a set of vectors is said to constitute an *orthonormal basis*, so named because each basis vector is a unit (or normal) vector, that is it has magnitude one, and also, by construction, each basis vector is perpendicular (or orthogonal) to each of the other two basis vectors.

2.4 Vector algebra and components

Given the vector **a**, as in equation (2.2), and a scalar λ, it follows from equation (1.2) (p. 8) that

$$\lambda\mathbf{a} = \lambda(a_1\mathbf{i} + a_2\mathbf{j} + a_3\mathbf{k}) = (\lambda a_1)\mathbf{i} + (\lambda a_2)\mathbf{j} + (\lambda a_3)\mathbf{k}.$$

So the vector $\lambda\mathbf{a}$ has components $\lambda a_1, \lambda a_2, \lambda a_3$.

As the basis vectors **i**, **j**, **k** are mutually perpendicular, we see, from Fig. 2.4, that by successive applications of Pythagoras' theorem

$$|\mathbf{a}| = \sqrt{(a_1{}^2 + a_2{}^2 + a_3{}^2)}. \tag{2.3}$$

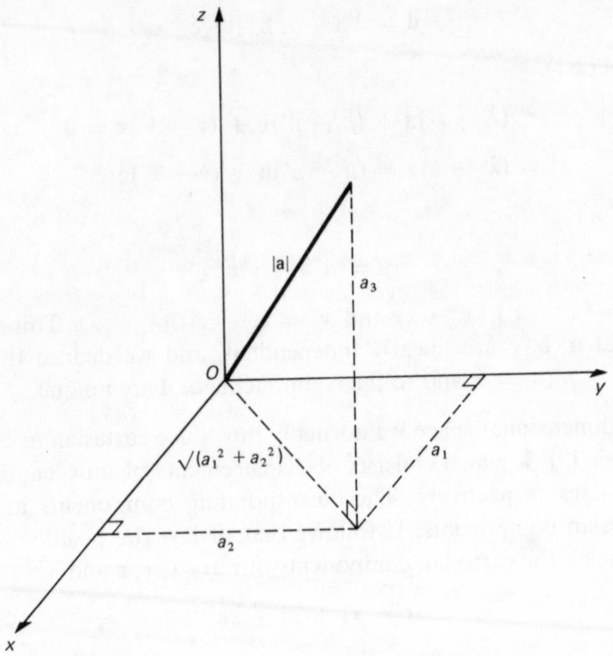

Fig. 2.4

Example 2 Calculate the magnitudes of the vectors

$$\mathbf{a} = 2\mathbf{i} - 3\mathbf{j} + \mathbf{k}, \quad \mathbf{b} = 3\mathbf{i} - \mathbf{j} + 5\mathbf{k}, \quad \mathbf{c} = -\mathbf{i} - 2\mathbf{j} - 4\mathbf{k}$$

and verify that $|\mathbf{b}|^2 = |\mathbf{a}|^2 + |\mathbf{c}|^2$.

From (2.3), $|\mathbf{a}|^2 = (2)^2 + (-3)^2 + (1)^2 = 4 + 9 + 1 = 14,$

$$|\mathbf{b}|^2 = (3)^2 + (-1)^2 + (5)^2 = 35,$$

$$|\mathbf{c}|^2 = (-1)^2 + (-2)^2 + (-4)^2 = 21,$$

$$\Rightarrow |\mathbf{a}| = \sqrt{14}, \quad |\mathbf{b}| = \sqrt{35}, \quad |\mathbf{c}| = \sqrt{21}$$

and $$|\mathbf{a}|^2 + |\mathbf{c}|^2 = 14 + 21 = 35 = |\mathbf{b}|^2.$$

Now consider the sum of two vectors $\mathbf{a} = a_1\mathbf{i} + a_2\mathbf{j} + a_3\mathbf{k}$ and $\mathbf{b} = b_1\mathbf{i} + b_2\mathbf{j} + b_3\mathbf{k}$. By the associative law of addition

$$\mathbf{a} + \mathbf{b} = (a_1\mathbf{i} + a_2\mathbf{j} + a_3\mathbf{k}) + (b_1\mathbf{i} + b_2\mathbf{j} + b_3\mathbf{k})$$

$$= (a_1 + b_1)\mathbf{i} + (a_2 + b_2)\mathbf{j} + (a_3 + b_3)\mathbf{k}.$$

Similarly, $$\mathbf{a} - \mathbf{b} = (a_1 - b_1)\mathbf{i} + (a_2 - b_2)\mathbf{j} + (a_3 - b_3)\mathbf{k}.$$

Then, since $\mathbf{0} = 0\mathbf{i} + 0\mathbf{j} + 0\mathbf{k}$, and the components of a vector are unique, we may deduce that, if $\mathbf{a} = \mathbf{b}$, then $a_1 = b_1$, $a_2 = b_2$ and $a_3 = b_3$.

Example 3 Show that, for the vectors in Example 2,

$$\mathbf{a} = \mathbf{b} + \mathbf{c}$$

and deduce that **a** and **c** are perpendicular.

Now
$$\mathbf{b} + \mathbf{c} = (3\mathbf{i} - \mathbf{j} + 5\mathbf{k}) + (-\mathbf{i} - 2\mathbf{j} - 4\mathbf{k})$$
$$= (2\mathbf{i} - 3\mathbf{j} + \mathbf{k}) = \mathbf{a}.$$

Hence the vectors **a**, **b** and **c** may be represented in magnitude and direction by the sides of a triangle. However, from the result $|\mathbf{b}|^2 = |\mathbf{a}|^2 + |\mathbf{c}|^2$, proved in Example 2, we then see that this triangle is right-angled with the side **b** being the hypotenuse.

Example 4 Given the vectors $\mathbf{a} = 2\mathbf{i} + 3\mathbf{j} - \mathbf{k}$, $\mathbf{b} = \mathbf{i} + 3\mathbf{j} - 2\mathbf{k}$, $\mathbf{c} = \mathbf{i} + \mathbf{j} - \mathbf{k}$ and $\mathbf{d} = \mathbf{i} + 3\mathbf{j} + 2\mathbf{k}$, express **d** as a linear combination of **a**, **b** and **c**.
Explain what the result tells you about **a**, **b** and **c**.

Let
$$\mathbf{d} = \alpha\mathbf{a} + \beta\mathbf{b} + \gamma\mathbf{c},$$

so that

$$\mathbf{i} + 3\mathbf{j} + 2\mathbf{k} = \alpha(2\mathbf{i} + 3\mathbf{j} - \mathbf{k}) + \beta(\mathbf{i} + 3\mathbf{j} - 2\mathbf{k}) + \gamma(\mathbf{i} + \mathbf{j} - \mathbf{k})$$
$$= (2\alpha + \beta + \gamma)\mathbf{i} + (3\alpha + 3\beta + \gamma)\mathbf{j} - (\alpha + 2\beta + \gamma)\mathbf{k}.$$

Equating corresponding components

$$\Rightarrow 1 = 2\alpha + \beta + \gamma, \quad 3 = 3\alpha + 3\beta + \gamma, \quad -2 = \alpha + 2\beta + \gamma$$

and straightforward algebra shows that these equations have the unique solution $\alpha = 8/3$, $\beta = -1/3$, $\gamma = -4$.
Thus

$$\mathbf{d} = \frac{8}{3}\mathbf{i} - \frac{1}{3}\mathbf{j} - 4\mathbf{k}.$$

As the equations for α, β, γ have a unique solution, we deduce that the vectors $\{\mathbf{a}, \mathbf{b}, \mathbf{c}\}$ form a linearly independent set. It is a useful exercise to prove the linear independence of **a**, **b**, **c** by showing that if $\lambda\mathbf{a} + \mu\mathbf{b} + \nu\mathbf{c} = \mathbf{0}$ then λ, μ and ν must all be zero. Consequently, it is not possible to express any one vector in the set $\{\mathbf{a}, \mathbf{b}, \mathbf{c}\}$ as a linear combination of the remaining two vectors.

2.5 Direction cosines

Let $\mathbf{a} = a_1\mathbf{i} + a_2\mathbf{j} + a_3\mathbf{k}$ be represented by the directed line $\overrightarrow{OA}$ so that $|\mathbf{a}| = OA$ and the direction of **a** is the same as the direction of $\overrightarrow{OA}$. Also let $\theta_1, \theta_2, \theta_3$ be the angles which OA makes with Ox, Oy, Oz, respectively. Figure 2.5 shows a rectangular parallelepiped (cuboid) which has concurrent edges of lengths a_1, a_2, a_3 along the x-, y-, z-axes, respectively, and has OA as a diagonal.

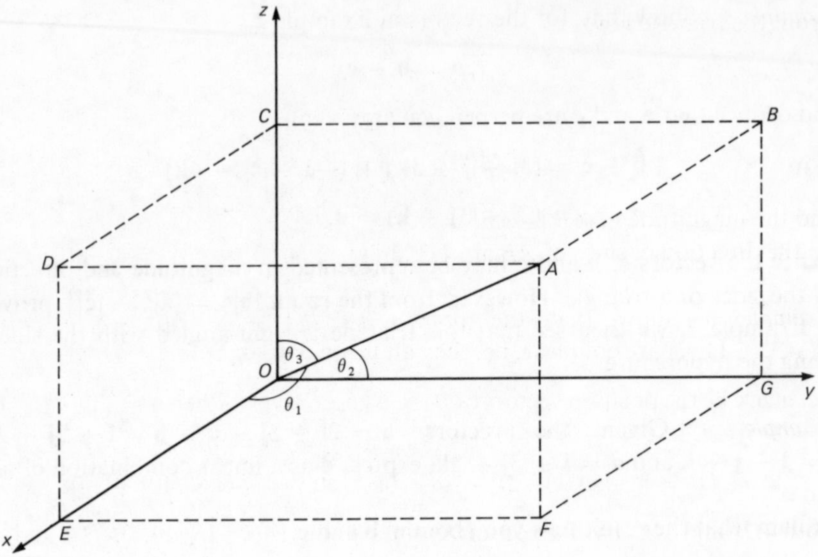

Fig. 2.5

The direction of $\overrightarrow{OA}$, and hence **a**, is sometimes specified, not by the angles $\theta_1, \theta_2, \theta_3$, but by their cosines $\cos\theta_1, \cos\theta_2, \cos\theta_3$. From Fig. 2.5 we see that

$$\cos\theta_1 = OE/OA = a_1/|\mathbf{a}|$$

and, similarly, $\cos\theta_2 = a_2/|\mathbf{a}|$ and $\cos\theta_3 = a_3/|\mathbf{a}|$. These three cosines, often denoted by l, m, n, respectively, are called the *direction cosines* of the vector **a**.

As $|\mathbf{a}|^2 = a_1{}^2 + a_2{}^2 + a_3{}^2$, dividing both sides by $|\mathbf{a}|^2$

$$\Rightarrow 1 = \cos^2\theta_1 + \cos^2\theta_2 + \cos^2\theta_3,$$

showing that the sum of the squares of the direction cosines is equal to unity.

Also $\qquad \mathbf{a} = a_1\mathbf{i} + a_2\mathbf{j} + a_3\mathbf{k}$

$$\Rightarrow \frac{\mathbf{a}}{|\mathbf{a}|} = \frac{a_1}{|\mathbf{a}|}\mathbf{i} + \frac{a_2}{|\mathbf{a}|}\mathbf{j} + \frac{a_3}{|\mathbf{a}|}\mathbf{k}$$

$$\Rightarrow \frac{\mathbf{a}}{|\mathbf{a}|} = \mathbf{i}\cos\theta_1 + \mathbf{j}\cos\theta_2 + \mathbf{k}\cos\theta_3,$$

so that the components of a unit vector are equal to the direction cosines of that vector.

Example 5 Given that points A and B have position vectors

$$\mathbf{a} = 4\mathbf{i} + 2\mathbf{j} + \mathbf{k} \quad \text{and} \quad \mathbf{b} = 5\mathbf{i} + 4\mathbf{j} - \mathbf{k}$$

respectively, find the direction cosines of $\overrightarrow{AB}$.

From the triangle law,

$$\overrightarrow{AB} = \mathbf{b} - \mathbf{a}$$

$$= \mathbf{i} + 2\mathbf{j} - 2\mathbf{k}.$$

$$\Rightarrow |\overrightarrow{AB}|^2 = 1 + 4 + 4 = 9$$

and the magnitude of $\overrightarrow{AB}$ is $|\overrightarrow{AB}| = 3$.
So the direction cosines of $\overrightarrow{AB}$ are 1/3, 2/3, −2/3.

Example 6 Show that the points $A(2, -1, -8)$, $B(0, -5, -2)$ and $C(-3, -11, 7)$ are collinear, i.e. they all lie on a straight line.

Let **a**, **b**, **c** be the position vectors of A, B, C respectively. Then $\mathbf{a} = 2\mathbf{i} - \mathbf{j} - 8\mathbf{k}$, $\mathbf{b} = -5\mathbf{j} - 2\mathbf{k}$, $\mathbf{c} = -3\mathbf{i} - 11\mathbf{j} + 7\mathbf{k}$.

$$\Rightarrow \overrightarrow{AB} = \mathbf{b} - \mathbf{a} = -2\mathbf{i} - 4\mathbf{j} + 6\mathbf{k}, \quad \overrightarrow{AC} = -5\mathbf{i} - 10\mathbf{j} + 15\mathbf{k}.$$

Both $\overrightarrow{AB}$ and $\overrightarrow{AC}$ have direction cosines $-1/\sqrt{14}$, $-2/\sqrt{14}$, $3/\sqrt{14}$ and, as the directed lines have a point A in common, we deduce that the points A, B, C are collinear.

Alternatively $\overrightarrow{AC} = \frac{5}{2}\overrightarrow{AB}$, so the displacements $\overrightarrow{AC}$ and $\overrightarrow{AB}$ are parallel. Hence, as they both contain the point A, the points A, B, C are collinear.

Exercise 2

1 Referred to axes Ox and Oy, the coordinates of the points P and Q are (3, 1) and (4, −3) respectively.
(a) Calculate the magnitude of the vector $\overrightarrow{OP} + 3\overrightarrow{OQ}$.
(b) Calculate the coordinates of the point R, if

$$\overrightarrow{OP} + \overrightarrow{OQ} = 2\overrightarrow{OR}.$$ (L)

2 Given that $\overrightarrow{OA} = \mathbf{a} = \mathbf{i} + 3\mathbf{j}$ and $\overrightarrow{OB} = \mathbf{b} = 3\mathbf{i} + 3\mathbf{j}$, plot on graph paper the points A and B. Plot also the points P and Q whose position vectors are **p** and **q** respectively, where $\mathbf{p} = \frac{1}{2}(\mathbf{a} + \mathbf{b})$ and $\mathbf{q} = \frac{1}{3}(\mathbf{a} + \mathbf{b})$. Describe the positions of P and Q.

3 The points A and B have position vectors $6\mathbf{i} - 5\mathbf{j}$ and $4\mathbf{i} - 3\mathbf{j}$ respectively. Show that the mid-point M of AB is collinear with the two points X, Y whose position vectors are $2\mathbf{i} - 6\mathbf{j}$ and $11\mathbf{i}$ respectively. (L)

4 The points A, B, C in the plane of the coordinate axes Ox, Oy have coordinates (1, 2), (4, 3), (3, −1) respectively, and D is the mid-point of the line BC. Express the vectors $\overrightarrow{BA}$ and $\overrightarrow{CA}$ in terms of **i** and **j**, and verify that

$$\overrightarrow{BA} + \overrightarrow{CA} = 2\overrightarrow{DA}.$$

5 The points A and B have position vectors $3\mathbf{i} + 2\mathbf{j}$ and $-\mathbf{i} + 4\mathbf{j}$ respectively, referred to O as origin. The point D has position vector $11\mathbf{i} - 2\mathbf{j}$ and is such that

$$\overrightarrow{OD} = m\overrightarrow{OA} + n\overrightarrow{OB}.$$

Calculate the values of m and n. (L)

6 The diagonals of a parallelogram $OABC$ intersect at M. Express each of the vectors (a) $\overrightarrow{OB}$, (b) $\overrightarrow{CA}$, (c) $\overrightarrow{OM}$ in terms of $\overrightarrow{OA}$ and $\overrightarrow{OC}$.

The coordinates of A and C referred to axes Ox and Oy are (3, 1) and (2, 4) re-

spectively. If D is the point which is such that $2\overrightarrow{OD} = 3\overrightarrow{CA}$, calculate
(d) the coordinates of D,
(e) the magnitude of the vector $(\overrightarrow{OD} + \overrightarrow{OM})$ and the acute angle made by this vector with the x-axis. (L)

7 Particles of mass 2g, 3g and 4g are situated at the points with position vectors $2\mathbf{i} - \mathbf{j}$, $-\mathbf{i} + 4\mathbf{j}$ and $3\mathbf{i} + 2\mathbf{j}$ respectively. Find the position vector of the centre of mass of these three particles.

8 Two points A and B have position vectors $\mathbf{a}$ and $\mathbf{b}$ respectively, relative to an origin O. The point C lies between A and B on the line AB and is such that $AC : CB = 1 : 2$. Calculate the position vector of the point C in terms of $\mathbf{a}$ and $\mathbf{b}$.

The point D lies in the plane of O, A and B and is such that $\overrightarrow{AD} = k.\overrightarrow{OB}$ and $\overrightarrow{OD} = h.\overrightarrow{OC}$. Write down two expressions for the position vector of D, one in terms of k and the other in terms of h. Use these expressions to find values for k and h. (L)

9 In the parallelogram $OPQR$, the position vectors of P and R with respect to O are $6\mathbf{i} + \mathbf{j}$ and $3\mathbf{i} + 4\mathbf{j}$ respectively. M and N are the mid-points of the sides PQ and QR respectively. Find
(a) the unit vector in the direction of $\overrightarrow{MN}$,
(b) the magnitude of the vector $\overrightarrow{OM}$. (L)

10 With O as origin the position vectors of the points A and B are $21\mathbf{j}$ and $8\mathbf{i} - 6\mathbf{j}$ respectively.
(a) Find the vector $\overrightarrow{OC}$, where C is the fourth vertex of the parallelogram $OACB$.
(b) Find the position vector of the point D in AB which is such that $AD : DB = 3 : 2$.
(c) Show that $3\overrightarrow{OC} + 5\overrightarrow{DO} = \overrightarrow{OA}$. (L)

11 Two unit vectors $\mathbf{a}$ and $\mathbf{b}$ are such that the direction of $\mathbf{b}$ makes an angle 60° with the direction of $\mathbf{a}$. Given that $\mathbf{v}_1 = 2\mathbf{a} - \mathbf{b}$, $\mathbf{v}_2 = -\mathbf{a} + 3\mathbf{b}$, $\mathbf{v}_3 = 3\mathbf{a} + \mathbf{b}$,
(i) find $\mathbf{v}_3$ in terms of $\mathbf{v}_1$ and $\mathbf{v}_2$,
(ii) prove that the magnitude of $\mathbf{v}_1 + 2\mathbf{v}_2 + 2\mathbf{v}_3$ is $\sqrt{127}$.

12 Prove that, if $\mathbf{a}$ and $\mathbf{b}$ are any two non-parallel, non-zero vectors lying in a plane, then any other vector $\mathbf{p}$ lying in this plane can be expressed in the form

$$\mathbf{p} = \lambda\mathbf{a} + \mu\mathbf{b},$$

where λ and μ are scalars.

Given that $\mathbf{a} = 2\mathbf{i} - 3\mathbf{j}$, $\mathbf{b} = -\mathbf{i} + 4\mathbf{j}$, $\mathbf{p} = 4\mathbf{i} - \mathbf{j}$, find λ and μ.

13 The position vectors of the points A, B, C, referred to an origin O in the plane ABC, are given by the unit vectors $\mathbf{a}$, $\mathbf{b}$, $\mathbf{c}$. Prove that

$$p\mathbf{a} + q\mathbf{b} + r\mathbf{c} = \mathbf{0},$$

where $\mathbf{0}$ is the null vector and p, q, r are scalar constants.

Describe the triangle ABC when
(i) $\mathbf{a} + \mathbf{b} + \mathbf{c} = \mathbf{0}$, (ii) $\mathbf{b} + \mathbf{c} = 0$.
Given that the line AO meets the line BC at the point K, prove that $OK = |p/(q + r)|$.

14 Given that the points A and B have position vectors $(2\mathbf{i} + 4\mathbf{j} + 7\mathbf{k})$ and $(-4\mathbf{i} + \mathbf{j} + \mathbf{k})$ respectively, find the position vector of the point P on AB which is such that $\overrightarrow{AP} = 2\overrightarrow{PB}$. (L)

15 Three vectors $\mathbf{a}$, $\mathbf{b}$, $\mathbf{c}$ lie in a plane such that $\mathbf{a} = 3\mathbf{i} - 2\mathbf{j}$, $\mathbf{b} = -2\mathbf{i} + \mathbf{j}$, $\mathbf{c} = 7\mathbf{i} - 4\mathbf{j}$. Find the resolution of each vector with respect to the basis formed by the other two.

16 Given the three vectors $\mathbf{a} = \mathbf{i} + 2\mathbf{j}$, $\mathbf{b} = -\mathbf{i} + 7\mathbf{j}$, $\mathbf{c} = 3\mathbf{i} - \mathbf{j}$, find the resolution of the vector $\mathbf{p} = \mathbf{a} + \mathbf{b} + \mathbf{c}$ with respect to the basis $\mathbf{a}$, $\mathbf{b}$.

17 Taking as basis the vectors $\overrightarrow{AB} = \mathbf{b}$ and $\overrightarrow{AC} = \mathbf{c}$ which coincide with two sides of a triangle ABC, find, as linear combinations of the base vectors, the vector representing the medians of the triangle.

18 Two points P and Q have position vectors $3\mathbf{i} - \mathbf{j} + 2\mathbf{k}$ and $\mathbf{i} - \mathbf{j} - 2\mathbf{k}$ with respect to an origin O. Calculate the length of PQ and show that the angle POQ is 90°.

19 Of the vectors $\mathbf{a} = 5\mathbf{i} + 6\mathbf{j} - 11\mathbf{k}$, $\mathbf{b} = 2\mathbf{i} - 2\mathbf{j} + 5\mathbf{k}$, $\mathbf{c} = 9\mathbf{i} + 2\mathbf{j} - \mathbf{k}$, $\mathbf{d} = -5\mathbf{i} - 14\mathbf{j} + 24\mathbf{k}$, show that **a**, **b**, **d** form a set of basis vectors. Express **c** in terms of this basis.

Given that **a**, **b**, **c** and **d** are the position vectors of points A, B, C and D respectively, show that the point $P(1, -2, 3)$ lies on AD and find the ratio $AP : AD$.

20 Given that $\overrightarrow{OA} = 2\mathbf{i} - 3\mathbf{j} + 4\mathbf{k}$, $\overrightarrow{OB} = 8\mathbf{i} + 6\mathbf{k}$, find the direction cosines of the line AB.

3 The scalar product

3.1 Scalar product of two vectors

In any triangle ABC, the cosine rule states that

$$c^2 = a^2 + b^2 - 2ab\cos\theta, \tag{3.1}$$

where a, b, c are the lengths of the sides BC, CA, AB respectively and θ is the angle ACB. If $\mathbf{a}$, $\mathbf{b}$, $\mathbf{c}$ are vectors represented by the displacements $\overrightarrow{CB}$, $\overrightarrow{CA}$, $\overrightarrow{AB}$ respectively, equation (3.1) may be written in the form

$$|\mathbf{c}|^2 = |\mathbf{a}|^2 + |\mathbf{b}|^2 - 2|\mathbf{a}|\,|\mathbf{b}|\cos\theta,$$

where θ is the angle between the vectors $\mathbf{a}$ and $\mathbf{b}$.

When using vectors in geometry or dynamics, the particular combination $|\mathbf{a}|\,|\mathbf{b}|\cos\theta$ of two vectors occurs frequently, and because of this we call it the scalar product of the two vectors.

Given any two vectors $\mathbf{a}$ and $\mathbf{b}$, the *scalar product* of $\mathbf{a}$ and $\mathbf{b}$ is denoted by $\mathbf{a}\,.\,\mathbf{b}$ and is defined to be the scalar equal to $|\mathbf{a}|\,|\mathbf{b}|\cos\theta$, where θ is the angle between the vectors.

So
$$\mathbf{a}\,.\,\mathbf{b} = |\mathbf{a}|\,|\mathbf{b}|\cos\theta$$

and the notation shows why the name *dot product* is used by many people.

3.2 Properties of the scalar product

If $\mathbf{a}$ and $\mathbf{b}$ are represented by directed lines $\overrightarrow{OA}$ and $\overrightarrow{OB}$, the expression $OA\cos\theta$ is the *projection* of the line OA onto the line OB. That is, in Fig. 3.1, the projection $OA\cos\theta = ON$, where N is the foot of the perpendicular from A onto OB. Consequently we see that

$$\begin{aligned}\mathbf{a}\,.\,\mathbf{b} &= |\mathbf{a}| \times (\text{the projection of } \mathbf{b} \text{ onto } \mathbf{a})\\ &= |\mathbf{b}| \times (\text{the projection of } \mathbf{a} \text{ onto } \mathbf{b}).\end{aligned}$$

It follows from the symmetry of the definition that, as $|\mathbf{a}|$, $|\mathbf{b}|$ and $\cos\theta$ are real numbers, forming the scalar product of two vectors is a commutative operation on those vectors, i.e.,

$$\mathbf{a}\,.\,\mathbf{b} = \mathbf{b}\,.\,\mathbf{a}.$$

When the two vectors are parallel, $\cos\theta = 1$ and $\mathbf{a}\,.\,\mathbf{b} = |\mathbf{a}|\,|\mathbf{b}|$. Consequently, $\mathbf{a}\,.\,\mathbf{a} = |\mathbf{a}|^2$ (sometimes written $\mathbf{a}^2$, or a^2). Similarly, when the

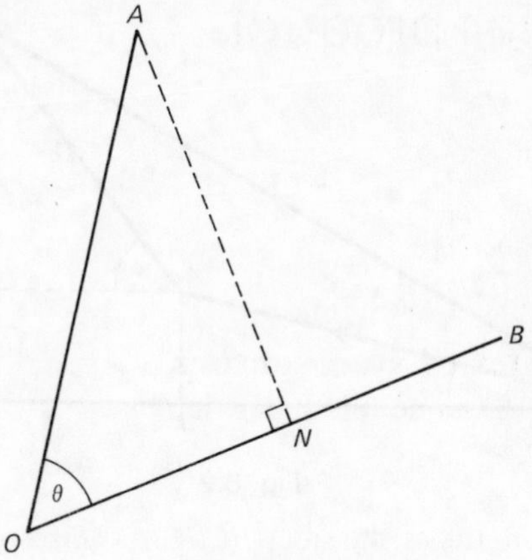

Fig. 3.1

vectors are perpendicular $\cos\theta = 0$ and $\mathbf{a}\,.\,\mathbf{b} = 0$. However, it should be noted that if $\mathbf{a}\,.\,\mathbf{b} = 0$ there are three possibilities: either $\mathbf{a} = \mathbf{0}$, or $\mathbf{b} = \mathbf{0}$ or $\mathbf{a}$ and $\mathbf{b}$ are perpendicular.

Example 1 Given two vectors $\mathbf{a}$ and $\mathbf{b}$, with $\mathbf{a} \neq \mathbf{0}$, describe the most general vector $\mathbf{x}$ satisfying the equation

$$\mathbf{a}\,.\,(\mathbf{x} - \mathbf{b}) = 0.$$

As $\mathbf{a} \neq \mathbf{0}$ there are two possibilities: either $\mathbf{x} = \mathbf{b}$ or $\mathbf{a}$ is perpendicular to $(\mathbf{x} - \mathbf{b})$. Both of these cases are incorporated in the general expression

$$\mathbf{x} = \mathbf{b} + \mathbf{c},$$

where $\mathbf{c}$ is an arbitrary (possibly zero) vector perpendicular to the vector $\mathbf{a}$, i.e., $\mathbf{c}$ is any vector such that $\mathbf{a}\,.\,\mathbf{c} = 0$.

The scalar product is also distributive in the sense that, for any vectors $\mathbf{a}$, $\mathbf{b}$, $\mathbf{c}$,

$$\mathbf{a}\,.\,(\mathbf{b} + \mathbf{c}) = \mathbf{a}\,.\,\mathbf{b} + \mathbf{a}\,.\,\mathbf{c}. \tag{3.2}$$

To prove this result in the case when all the scalar products are positive, we use the diagram in Fig. 3.2.

$$\begin{aligned}
\mathbf{a}\,.\,(\mathbf{b} + \mathbf{c}) &= |\mathbf{a}|\,|\mathbf{b} + \mathbf{c}| \cos\psi \\
&= |\mathbf{a}|\,(OM) \\
&= |\mathbf{a}|\,(OL + LM) \\
&= |\mathbf{a}|\,(OL) + |\mathbf{a}|\,(AN) \\
&= |\mathbf{a}|\,|\mathbf{b}| \cos\theta + |\mathbf{a}|\,|\mathbf{c}| \cos\phi \\
&= \mathbf{a}\,.\,\mathbf{b} + \mathbf{a}\,.\,\mathbf{c}.
\end{aligned}$$

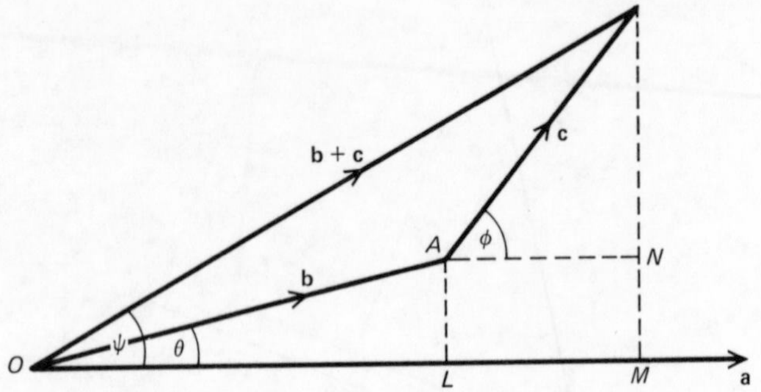

Fig. 3.2

Similar proofs may be constructed to cover the other cases, e.g. when $\cos \psi < 0$.

Example 2 Prove that the mid-point of the hypotenuse of a right-angled triangle is equidistant from all three vertices.

Let the triangle have vertices O, A, B and let the angle at O be the right-angle. (The reader is advised to draw a sketch.) Then, if $\mathbf{a} = \overrightarrow{OA}$ and $\mathbf{b} = \overrightarrow{OB}$, the mid-point M of the hypotenuse AB is such that $\overrightarrow{OM} = \frac{1}{2}(\mathbf{a} + \mathbf{b})$.

Then
$$OM^2 = |\overrightarrow{OM}|^2 = \tfrac{1}{4}(\mathbf{a} + \mathbf{b}) \cdot (\mathbf{a} + \mathbf{b})$$
$$= \tfrac{1}{4}(|\mathbf{a}|^2 + |\mathbf{b}|^2 + 2\mathbf{a} \cdot \mathbf{b}).$$

However, as $\mathbf{a}$ and $\mathbf{b}$ are perpendicular, $\mathbf{a} \cdot \mathbf{b} = 0$, and so
$$OM^2 = \tfrac{1}{4}(|\mathbf{a}|^2 + |\mathbf{b}|^2).$$

Similarly,
$$AM^2 = |\overrightarrow{AM}|^2$$
$$= |\tfrac{1}{2}(\mathbf{b} - \mathbf{a})|^2$$
$$= \tfrac{1}{4}(|\mathbf{b}|^2 + |\mathbf{a}|^2 - 2\mathbf{a} \cdot \mathbf{b})$$
$$= \tfrac{1}{4}(|\mathbf{b}|^2 + |\mathbf{a}|^2).$$

and
$$BM^2 = |\overrightarrow{BM}|^2$$
$$= |\tfrac{1}{2}(\mathbf{a} - \mathbf{b})|^2$$
$$= \tfrac{1}{4}(|\mathbf{a}|^2 + |\mathbf{b}|^2 - 2\mathbf{a} \cdot \mathbf{b})$$
$$= \tfrac{1}{4}(|\mathbf{a}|^2 + |\mathbf{b}|^2).$$

Hence $OM = AM = BM$.

Example 3 Prove that the altitudes of a triangle are concurrent.

In Fig. 3.3 the altitudes through the vertices A and B meet in a point O and $\mathbf{a}$, $\mathbf{b}$, $\mathbf{c}$ are the position vectors of the vertices A, B, C respectively relative to O.

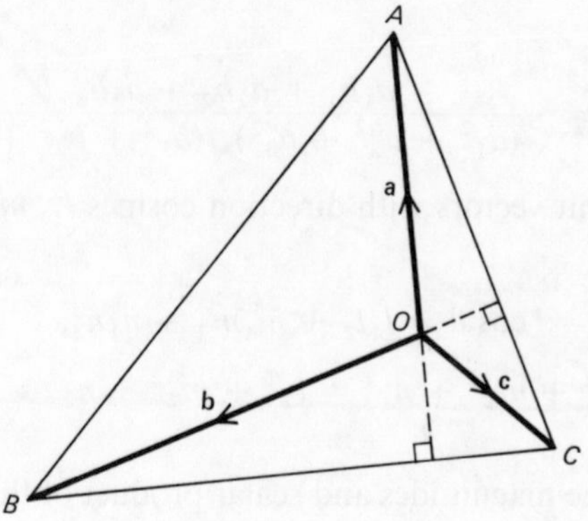

Fig. 3.3

Since, by construction, $\overrightarrow{OA}$ and $\overrightarrow{BC} = \mathbf{c} - \mathbf{b}$ are perpendicular, we have $\mathbf{a}.(\mathbf{c} - \mathbf{b}) = 0$, and similarly, as $\overrightarrow{OB}$ and $\overrightarrow{AC}$ are perpendicular, we have

$$\mathbf{b}.(\mathbf{c} - \mathbf{a}) = 0.$$

Thus

$$\mathbf{a}.\mathbf{c} = \mathbf{a}.\mathbf{b} \quad \text{and} \quad \mathbf{b}.\mathbf{c} = \mathbf{b}.\mathbf{a}.$$

$$\Rightarrow \mathbf{a}.\mathbf{c} - \mathbf{b}.\mathbf{c} = 0$$

$$\Rightarrow \mathbf{c}.(\mathbf{b} - \mathbf{a}) = 0.$$

Assuming A, B, C to be distinct points, so $\mathbf{a} \neq \mathbf{b}$, we have either (i) $\mathbf{c} = \mathbf{0}$, which means that O and C coincide and the triangle has a right-angle at C so that the altitudes all intersect at C, or (ii) $\mathbf{c}$ and $\mathbf{b} - \mathbf{a} = \overrightarrow{AB}$ are perpendicular, which means that OC is an altitude and so all three altitudes are concurrent at O.

As the cartesian basis vectors $\mathbf{i}$, $\mathbf{j}$, $\mathbf{k}$ are mutually orthogonal, we have

$$\mathbf{j}.\mathbf{k} = \mathbf{k}.\mathbf{i} = \mathbf{i}.\mathbf{j} = 0 \tag{3.3}$$

and, as they are unit vectors,

$$\mathbf{i}.\mathbf{i} = \mathbf{j}.\mathbf{j} = \mathbf{k}.\mathbf{k} = 1. \tag{3.4}$$

Consequently if $\mathbf{a} = a_1\mathbf{i} + a_2\mathbf{j} + a_3\mathbf{k}$ and $\mathbf{b} = b_1\mathbf{i} + b_2\mathbf{j} + b_3\mathbf{k}$ we have, from (3.2), (3.3) and (3.4),

$$\mathbf{a}.\mathbf{b} = a_1b_1 + a_2b_2 + a_3b_3 \tag{3.5}$$

and

$$\mathbf{a}.\mathbf{a} = |a|^2 = a_1{}^2 + a_2{}^2 + a_3{}^2. \tag{3.6}$$

Equation (3.6) is of course equivalent to equation (2.3) (p. 17), which was obtained geometrically.

From the definition of scalar product we can now see that the angle between

a and **b** is given by

$$\cos\theta = \frac{a_1b_1 + a_2b_2 + a_3b_3}{\sqrt{(a_1{}^2 + a_2{}^2 + a_3{}^2)}\sqrt{(b_1{}^2 + b_2{}^2 + b_3{}^2)}},$$

or, if **a** and **b** are unit vectors with direction cosines l_1, m_1, n_1 and l_2, m_2, n_2 respectively,

$$\cos\theta = l_1l_2 + m_1m_2 + n_1n_2,$$

since $$l_1{}^2 + m_1{}^2 + n_1{}^2 = l_2{}^2 + m_2{}^2 + n_2{}^2 = 1.$$

Example 4 Find the magnitudes and scalar product of the vectors

$$\mathbf{a} = \mathbf{i} + 3\mathbf{j} + \mathbf{k} \quad \text{and} \quad \mathbf{b} = 3\mathbf{i} - \mathbf{j} - \mathbf{k}.$$

Hence find the angle between the vectors.

Clearly $|\mathbf{a}|^2 = 1^2 + 3^2 + 1^2 = 11, \quad |\mathbf{b}|^2 = 3^2 + (-1)^2 + (-1)^2 = 11.$

Also $\mathbf{a}\,.\,\mathbf{b} = (1)(3) + (3)(-1) + (1)(-1) = -1.$

Thus, if θ is the angle between **a** and **b**,

$$\cos\theta = \frac{-1}{\sqrt{11}\sqrt{11}} = -\frac{1}{11}$$

$$\Rightarrow \theta \approx 1{\cdot}66 \text{ radians} \approx 95{\cdot}2°.$$

Example 5 Find a vector **c** which is perpendicular to both of the vectors

$$\mathbf{a} = \mathbf{i} + 3\mathbf{j} - \mathbf{k} \quad \text{and} \quad \mathbf{b} = 3\mathbf{i} - \mathbf{j} - \mathbf{k}.$$

We require $\mathbf{a}\,.\,\mathbf{c} = 0$ and $\mathbf{b}\,.\,\mathbf{c} = 0$.
So if $\mathbf{c} = c_1\mathbf{i} + c_2\mathbf{j} + c_3\mathbf{k}$, we have

$$c_1 + 3c_2 - c_3 = 0 \quad \text{and} \quad 3c_1 - c_2 - c_3 = 0.$$

These have solution $c_1 = 2c_2$, $c_3 = 5c_2$ and so the vector $\mathbf{c} = c_2(2\mathbf{i} + \mathbf{j} + 5\mathbf{k})$ is perpendicular to both **a** and **b** for all values of c_2.

A vector is easily expressed as the sum of three mutually orthogonal vectors. For example, if the vectors are **i**, **j** and **k**, any vector **a** has the form $\mathbf{a} = a_1\mathbf{i} + a_2\mathbf{j} + a_3\mathbf{k}$ and, taking scalar products with **i**, **j**, **k**,

$$a_1 = \mathbf{a}\,.\,\mathbf{i}, \quad a_2 = \mathbf{a}\,.\,\mathbf{j}, \quad a_3 = \mathbf{a}\,.\,\mathbf{k}$$

$$\Rightarrow \mathbf{a} = (\mathbf{a}\,.\,\mathbf{i})\mathbf{i} + (\mathbf{a}\,.\,\mathbf{j})\mathbf{j} + (\mathbf{a}\,.\,\mathbf{k})\mathbf{k}.$$

However, in problems of kinematics and dynamics we sometimes need to express a given vector **a** as the sum of two vectors, one parallel to a given vector **b** and the other, **c**, perpendicular to **b** (see Fig. 3.4). That is, $\mathbf{a} = \lambda\mathbf{b} + \mathbf{c}$, where $\lambda\mathbf{b}\,.\,\mathbf{c} = 0$. Clearly $\mathbf{c} = \mathbf{a} - \lambda\mathbf{b}$ and λ is determined from

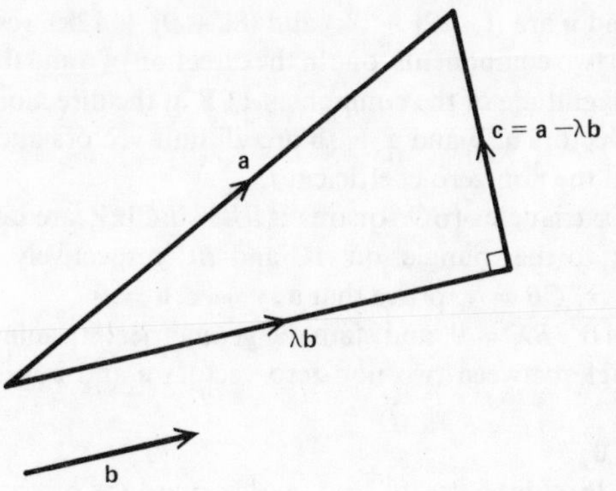

Fig. 3.4

$$\lambda\mathbf{b}\,.(\mathbf{a}-\lambda\mathbf{b})=0,$$

$$\Rightarrow \lambda\mathbf{a}\,.\,\mathbf{b}-\lambda^2|\mathbf{b}|^2=0,$$

$\Rightarrow \lambda = \mathbf{a}\,.\,\mathbf{b}/|\mathbf{b}|^2$, since $\lambda = 0$ corresponds to **a** and **b** being perpendicular.

So, if $\hat{\mathbf{b}} = \mathbf{b}/|\mathbf{b}|$ is the unit vector parallel to **b**, the vector **a** is equal to the sum of the two vectors

$$(\mathbf{a}\,.\,\hat{\mathbf{b}})\hat{\mathbf{b}} \quad \text{and} \quad \mathbf{a}-(\mathbf{a}\,.\,\hat{\mathbf{b}})\hat{\mathbf{b}}.$$

The first of these vectors is parallel to **b** and the second is perpendicular to **b**.

Example 6 Express the vector $\mathbf{i}+3\mathbf{j}+\mathbf{k}$ as the sum of two vectors, one parallel to and one perpendicular to $\mathbf{i}-2\mathbf{j}+2\mathbf{k}$.

Let $\mathbf{a}=\mathbf{i}+3\mathbf{j}+\mathbf{k}$ and $\mathbf{b}=\mathbf{i}-2\mathbf{j}+2\mathbf{k}$. Then

$$\hat{\mathbf{b}}=\tfrac{1}{3}(\mathbf{i}-2\mathbf{j}+2\mathbf{k})$$

$$\Rightarrow \mathbf{a}\,.\,\hat{\mathbf{b}}=\tfrac{1}{3}(1-6+2)=-1.$$

The required vectors are thus

$$(\mathbf{a}\,.\,\hat{\mathbf{b}})\hat{\mathbf{b}}=\tfrac{1}{3}(-\mathbf{i}+2\mathbf{j}-2\mathbf{k})$$

and $$\mathbf{a}-(\mathbf{a}\,.\,\hat{\mathbf{b}})\hat{\mathbf{b}}=\tfrac{1}{3}(4\mathbf{i}+7\mathbf{j}+5\mathbf{k}).$$

Exercise 3

1 Find the value of λ for which the vectors $(2\mathbf{i}-3\mathbf{j}+\mathbf{k})$ and $(3\mathbf{i}+6\mathbf{j}+\lambda\mathbf{k})$ are perpendicular. (L)

2 Given the vectors $\mathbf{u}=3\mathbf{i}+2\mathbf{j}$ and $\mathbf{v}=2\mathbf{i}+\lambda\mathbf{j}$, determine the values of λ so that
(a) **u** and **v** are at right angles,
(b) **u** and **v** are parallel,
(c) the acute angle between **u** and **v** is $\pi/4$. (L)

3 The angle between the vectors $(\mathbf{i}+\mathbf{j})$ and $(\mathbf{i}+\mathbf{j}+p\mathbf{k})$ is $\pi/4$. Find the possible values of p. (L)

4 The vectors **F** and **u** are (**i** − 2**j** + 3**k**) and (8**i** + 9**j** + 12**k**), respectively. The vector **F** is resolved into two components, one in the direction of **u** and the other perpendicular to **u**. Find the magnitude of the component of **F** in the direction of **u**. (L)

5 Given that the vectors **a**, **b** and **a** − t**b** are all unit vectors and the angle between **a** and **b** is $\pi/4$, find the non-zero coefficient t.

6 In the plane of a triangle ABC squares $ACXY$, $BCWZ$ are described, in the order given, externally to the triangle, on AC and BC respectively. Given that CX = **b**, CA = **a**, CW = **x**, CB = **y**, prove that **a**.**y** + **x**.**b** = 0.

Deduce that $\overrightarrow{AW}.\overrightarrow{BX} = 0$, and state the geometrical meaning of this result.

7 (a) State the angle between two non-zero vectors **a** and **b** in each of the following cases.

(i) $2\mathbf{a} + 3\mathbf{b} = \mathbf{0}$;

(ii) $|\mathbf{a}| = |\mathbf{b}| = |\mathbf{a} + \mathbf{b}|$;

(iii) $\mathbf{a}.(\mathbf{a} + \mathbf{b}) = 0$ and $|\mathbf{b}| = 2|\mathbf{a}|$.

(b) If **c** is a vector coplanar with two non-parallel vectors **p** and **q**, show that **c** can be expressed uniquely in the form

$$\mathbf{c} = \lambda\mathbf{p} + \mu\mathbf{q},$$

where λ and μ are constants.

Given that **p** = 2**i** + 3**j** and **q** = 3**i** + 4**j**, find the numbers λ and μ for the vector **c** which is of magnitude 10 units and makes the angle arc tan $\frac{3}{4}$ with the positive Ox axis in the first quadrant.

8 A tetrahedron $OABC$ with vertex O at the origin is such that $\overrightarrow{OA}$ = **a**, $\overrightarrow{OB}$ = **b** and $\overrightarrow{OC}$ = **c**. Show that the line segments joining the mid-points of opposite edges bisect one another. Given that two pairs of opposite edges are perpendicular, prove that **b**.**c** = **c**.**a** = **a**.**b** and show that the third pair of opposite edges is also perpendicular.

Prove also that, in this case, $OA^2 + BC^2 = OB^2 + AC^2$. (L)

9 Given that $\overrightarrow{OA}$ = **i** + **j** and $\overrightarrow{OB}$ = 5**i** + 7**j**, find the position vectors of the other two vertices of the square of which A and B are one pair of opposite vertices.

10 The position vectors of points A and B with respect to an origin O are **a** and **b**, respectively. If $\triangle$ is the area of the triangle AOB, show that

$$4\triangle^2 = |\mathbf{a}|^2|\mathbf{b}|^2 - (\mathbf{a}.\mathbf{b})^2.$$

Given that **a** = **i** + 2**j** + 2**k** and **b** = 4**i** − 4**j** + 2**k**, find the area and the angles of the triangle AOB.

11 Find the cosines of the angles of the triangle with vertices A(**i** − **k**), B(3**i** + 2**j** + **k**), C(2**i** + **j** + 3**k**).

12 Show that

$$|\mathbf{p} + \mathbf{q}|^2 + |\mathbf{p} - \mathbf{q}|^2 = 2|\mathbf{p}|^2 + 2|\mathbf{q}|^2.$$

If $ABCD$ is a rectangle and O is any other point in its plane, show that $OA^2 + OC^2 = OB^2 + OD^2$.

13 The position vectors of A and B with respect to an origin O are **a** and **b**, respectively. Show that the position vector of any point P on AB is of the form $k\mathbf{a} + (1 - k)\mathbf{b}$.

Prove that $BP^2 = k^2(a^2 + b^2 - 2\mathbf{a}.\mathbf{b})$ where $a = |\mathbf{a}|$ and $b = |\mathbf{b}|$. Derive similar expressions for AP^2 and OP^2.

By taking $k = \frac{1}{2}$, deduce that $OA^2 + OB^2 = 2OP^2 + 2AP^2$, where P is the mid-point of AB.

14 The unit vectors **s**, **t** in the first quadrant of the x–y plane, have directions making angles $\frac{1}{4}\pi$, arc tan $\frac{3}{4}$ respectively with Ox. Given that the resolutes of **a** in the directions of **s** and the positive x-axis are $\sqrt{2}$, -1 respectively, and the resolutes of **b** in the directions of **s** and **t** are $\sqrt{2}$, -1 respectively, find the cartesian components of both **a** and **b**. Find the components in the directions of **s** and **t** of the vector **c** with resolutes 4, $3\sqrt{2}$ in the directions of **s** and **t** respectively.

15 *ABCD* is a tetrahedron whose vertices have position vectors **a**, **b**, **c**, **d** respectively. The edge *DA* is perpendicular to *BC*, and the edge *DB* is perpendicular to *CA*. Show vectorially that *DC* is perpendicular to *AB*.

Given that no two of $\overrightarrow{DA}$, $\overrightarrow{DB}$, $\overrightarrow{DC}$ are perpendicular, show, by taking *D* as origin, that

$$|DA|:|DB|:|DC| = \mathbf{q}.\mathbf{r}:\mathbf{r}.\mathbf{p}:\mathbf{p}.\mathbf{q},$$

where **p**, **q** and **r** are unit vectors in the directions of $\overrightarrow{DA}$, $\overrightarrow{DB}$ and $\overrightarrow{DC}$ respectively.

By taking $\mathbf{t} = \alpha\mathbf{p} + \beta\mathbf{q} + \gamma\mathbf{r}$, where α, β, γ are numbers to be evaluated, find a unit vector **t** perpendicular to the plane *ABC* when the angles *BDC* and *CDA* are each $\frac{1}{3}\pi$ and the angle *ADB* is arc cos $\frac{1}{3}$.

16 The angles between three non-zero vectors **a**, **b**, **c**, which are not necessarily coplanar, are α between **b** and **c**, β between **c** and **a**, γ between **a** and **b**. Vectors **u** and **v** are defined by

$$\mathbf{u} = (\mathbf{a}.\mathbf{c})\mathbf{b} - (\mathbf{a}.\mathbf{b})\mathbf{c},$$

$$\mathbf{v} = (\mathbf{a}.\mathbf{c})\mathbf{b} - (\mathbf{b}.\mathbf{c})\mathbf{a}.$$

Given that **u** and **v** are perpendicular, write down the value of the scalar product **u.v** and show that either **c** is perpendicular to **a** or else

$$\cos\beta = \cos\alpha\cos\gamma.$$

Hence show that, if **a**, **b**, **c** are coplanar vectors, then either **c** is perpendicular to **a** or else **b** is parallel to either **c** or **a**.

4 Applications of vectors

4.1 Notation

The results of the preceding chapters are now used in three different circumstances. The principal application to be discussed in this book is the geometry of real space but some elementary kinematical and physical applications are also introduced. As we frequently refer to the position vectors of points relative to an origin O, it is convenient to introduce the notation 'the point $A(\mathbf{a})$' to denote 'the point A whose position vector relative to the fixed origin O is $\mathbf{a}$'.

4.2 Geometrical applications

Here we consider only the geometry of straight lines and planes. A straight line is the shape taken by an infinitely long, taut, weightless string, whilst a plane may be identified with a flat table-top extending to infinity in all directions. Thus it is geometrically clear that two lines in real space do not, in general, meet and that two non-parallel planes will have a common line of intersection.

Equations of a straight line

First consider the straight line through the point $A(\mathbf{a})$ parallel to the vector $\mathbf{m}$ (see Fig. 4.1).

If $P(\mathbf{r})$ is any point on the line, the vector displacement $\overrightarrow{AP}$ is parallel to $\mathbf{m}$ and so there is some scalar λ such that $\overrightarrow{AP} = \lambda\mathbf{m}$.

Then $$\overrightarrow{OP} = \overrightarrow{OA} + \overrightarrow{AP}$$

$$\Rightarrow \mathbf{r} = \mathbf{a} + \lambda\mathbf{m}. \qquad (4.1)$$

Note that λ is positive for points P on one side of A and negative for points on the other side of A. Also for any point P on the line there is a unique value of λ and, conversely, for any value of λ there is a unique point P on the line. Consequently, when λ varies over all real values, the point whose position vector is given by (4.1) moves along the line through $A(\mathbf{a})$ parallel to $\mathbf{m}$. Thus (4.1) is called an equation of the line through $A(\mathbf{a})$ parallel to $\mathbf{m}$. Sometimes (4.1) is called the *vector parametric* form of the equation of the line.

If (x, y, z) and (a_1, a_2, a_3) are the cartesian coordinates of P and A, respectively, referred to axes through O and if $\mathbf{m}$ has cartesian components m_1, m_2, m_3, then equation (4.1)

$$\Rightarrow x\mathbf{i} + y\mathbf{j} + z\mathbf{k} = (a_1 + \lambda m_1)\mathbf{i} + (a_2 + \lambda m_2)\mathbf{j} + (a_3 + \lambda m_3)\mathbf{k}.$$

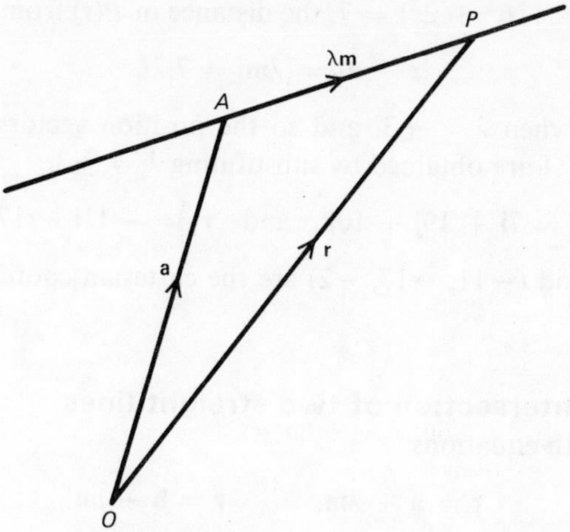

Fig. 4.1

Equating corresponding components,

$$(x - a_1):m_1 = (y - a_2):m_2 = (z - a_3):m_3,$$

which are the cartesian equations of the line passing through the point with coordinates (a_1, a_2, a_3) and with direction cosines proportional to m_1, m_2, m_3.

To find an equation of the straight line passing through the points $A(\mathbf{a})$ and $B(\mathbf{b})$, we note that $\overrightarrow{AB} = \mathbf{b} - \mathbf{a}$ is parallel to the line. Hence we may take $\mathbf{m} = \mathbf{b} - \mathbf{a}$ and a vector parametric equation of the line is

$$\mathbf{r} = \mathbf{a} + \lambda(\mathbf{b} - \mathbf{a})$$

or

$$\mathbf{r} = (1 - \lambda)\mathbf{a} + \lambda\mathbf{b}.$$

Example 1 Find the vector parametric equation and the cartesian equations of the straight line joining the points A and B, whose cartesian coordinates are $(-2, 1, 4)$ and $(1, 7, 6)$ respectively. Also, find two points on the line whose distances from A are each 21.

As $\mathbf{a} = -2\mathbf{i} + \mathbf{j} + 4\mathbf{k}$ and $\mathbf{b} = \mathbf{i} + 7\mathbf{j} + 6\mathbf{k}$, the line is parallel to the vector $\mathbf{m} = \mathbf{b} - \mathbf{a} = 3\mathbf{i} + 6\mathbf{j} + 2\mathbf{k}$.

Hence the vector parametric equation of the line is

$$\mathbf{r} = x\mathbf{i} + y\mathbf{j} + z\mathbf{k} = (-2\mathbf{i} + \mathbf{j} + 4\mathbf{k}) + \lambda(3\mathbf{i} + 6\mathbf{j} + 2\mathbf{k})$$

$$\Rightarrow (x + 2)\mathbf{i} + (y - 1)\mathbf{j} + (z - 4)\mathbf{k} = \lambda(3\mathbf{i} + 6\mathbf{j} + 2\mathbf{k}).$$

Equating components gives the cartesian equations

$$\frac{x + 2}{3} = \frac{y - 1}{6} = \frac{z - 4}{2}.$$

Since $|\mathbf{m}| = \sqrt{(3^2 + 6^2 + 2^2)} = 7$, the distance of $P(\mathbf{r})$ from $A(\mathbf{a})$ is

$$|\mathbf{r} - \mathbf{a}| = |\lambda\mathbf{m}| = 7|\lambda|.$$

Now $7|\lambda| = 21$ when $\lambda = \pm 3$, and so the position vectors $\mathbf{r}_1$, $\mathbf{r}_2$ of points distance 21 from A are obtained by substituting $\lambda = \pm 3$.

Thus $\qquad \mathbf{r}_1 = 7\mathbf{i} + 19\mathbf{j} + 10\mathbf{k}$ and $\mathbf{r}_2 = -11\mathbf{i} - 17\mathbf{j} - 2\mathbf{k}$,

and (7, 19, 10) and (−11, −17, −2) are the cartesian coordinates of the two points.

The point of intersection of two straight lines

The two lines with equations

$$\mathbf{r} = \mathbf{a} + \lambda\mathbf{m}, \qquad \mathbf{r} = \mathbf{b} + \mu\mathbf{n}$$

may or may not intersect. When they do intersect, there exists a point $P(\mathbf{r}_1)$ such that $\mathbf{r} = \mathbf{r}_1$ satisfies both equations. Consequently the corresponding values of λ and μ must satisfy

$$\mathbf{a} + \lambda\mathbf{m} = \mathbf{b} + \mu\mathbf{n}.$$

Expressing **a**, **b**, **m** and **n** in terms of their components and equating corresponding components gives three equations for the two scalar variables λ and μ. Then either

(i) there are no values of λ and μ which simultaneously satisfy all three equations, and this is the usual situation; in this case the lines do not intersect as there is no point $P(\mathbf{r}_1)$ common to both, i.e. the lines are *skew*, or

(ii) in special circumstances, there are particular values λ_1 and μ_1 which satisfy all three equations. This means that there is a point $P(\mathbf{r}_1)$ lying on both lines and $\mathbf{r}_1$ is obtained by substituting λ_1 (or μ_1) into the equation of the appropriate line.

Example 2 The three lines L_1, L_2 and L_3 are given by the equations

$$L_1 : \mathbf{r} = 7\mathbf{i} - 3\mathbf{j} + 3\mathbf{k} + \lambda(3\mathbf{i} - 2\mathbf{j} + \mathbf{k}),$$

$$L_2 : \mathbf{r} = 7\mathbf{i} - 2\mathbf{j} + 4\mathbf{k} + \mu(-2\mathbf{i} + \mathbf{j} - \mathbf{k}),$$

$$L_3 : \mathbf{r} = \mathbf{i} + \nu(\mathbf{j} - \mathbf{k}).$$

Show that L_1 and L_2 intersect and find the point of intersection. Show also that L_1 and L_3 do not intersect.

If L_1 and L_2 intersect, λ and μ must satisfy

$$7\mathbf{i} - 3\mathbf{j} + 3\mathbf{k} + \lambda(3\mathbf{i} - 2\mathbf{j} + \mathbf{k}) = 7\mathbf{i} - 2\mathbf{j} + 4\mathbf{k} + \mu(-2\mathbf{i} + \mathbf{j} - \mathbf{k})$$

and equating corresponding components

$$\Rightarrow 3\lambda + 2\mu = 0, \quad 2\lambda + \mu = -1, \quad \lambda + \mu = 1.$$

All three equations are satisfied by $\lambda = -2, \mu = 3$ and so L_1 and L_2 intersect. The position vector of the point of intersection, found by substituting $\lambda = -2$ in the equation for L_1 or $\mu = 3$ in the equation for L_2, is $\mathbf{i} + \mathbf{j} + \mathbf{k}$.

If L_1 and L_3 intersect, λ and ν must satisfy

$$7\mathbf{i} - 3\mathbf{j} + 3\mathbf{k} + \lambda(3\mathbf{i} - 2\mathbf{j} + \mathbf{k}) = \mathbf{i} + \nu(\mathbf{j} - \mathbf{k})$$

$$\Rightarrow 3\lambda = -6, \quad 2\lambda + \nu = -3, \quad \lambda + \nu = -3.$$

These equations are not consistent, for, while the first two give $\lambda = -2, \nu = 1$, these values do not satisfy the third equation. Hence, we conclude that L_1 and L_2 do not intersect.

The shortest distance between two straight lines

Consider the two lines L_1 and L_2 with equations

$$\mathbf{r} = \mathbf{a} + \lambda\mathbf{m}, \qquad \mathbf{r} = \mathbf{b} + \mu\mathbf{n}$$

respectively. Assume that the lines do not intersect, so that, if P lies on L_1 and Q lies on L_2, the distance PQ is never zero. In this case there is a minimum distance between points on the two lines and it occurs when the line PQ is perpendicular to both L_1 and L_2 (see Fig. 4.2).

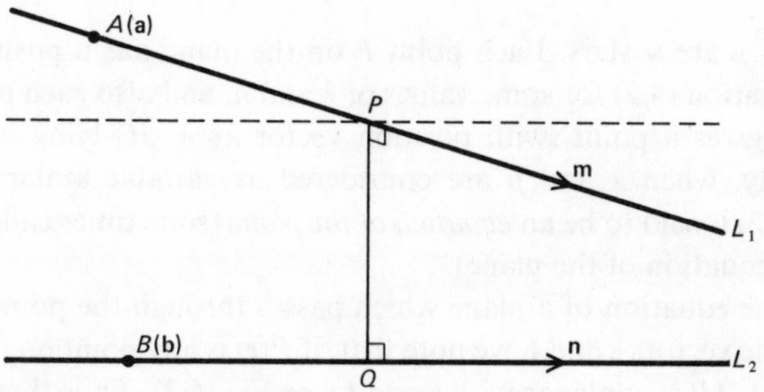

Fig. 4.2

To find a concise expression, in terms of **a**, **b**, **m** and **n**, for the minimum distance between L_1 and L_2 it is easiest to use the *vector product* of two vectors, which will not be introduced until chapter 6. However, for the moment we may proceed as follows in individual cases.

As PQ is perpendicular to both L_1 and L_2, we have

$$\overrightarrow{PQ}\cdot\mathbf{m} = \overrightarrow{PQ}\cdot\mathbf{n} = 0,$$

where $\overrightarrow{PQ} = (\mathbf{b} - \mathbf{a}) + (\mu\mathbf{n} - \lambda\mathbf{m})$. These equations may be solved to find the values of λ and μ corresponding to the points P and Q, and hence the position vectors of the points P and Q and so the minimum distance $h = |\overrightarrow{PQ}|$.

Example 3 Find the minimum distance between the lines

$$L_1 : \mathbf{r} = 3\mathbf{i} - 2\mathbf{j} - \mathbf{k} + \lambda(2\mathbf{i} - \mathbf{j} + 2\mathbf{k}) = \mathbf{a} + \lambda\mathbf{m},$$

$$L_2 : \mathbf{r} = 2\mathbf{i} + \mathbf{j} - 3\mathbf{k} + \mu(2\mathbf{i} - 3\mathbf{j} + 2\mathbf{k}) = \mathbf{b} + \mu\mathbf{n}.$$

$$\overrightarrow{PQ} = (\mathbf{b} - \mathbf{a}) + (\mu\mathbf{n} - \lambda\mathbf{m})$$

$$= (2\mu - 2\lambda - 1)\mathbf{i} + (-3\mu + \lambda + 3)\mathbf{j} + (2\mu - 2\lambda - 2)\mathbf{k}.$$

Then $\overrightarrow{PQ}.\mathbf{m} = 0 \Rightarrow 11\mu - 9\lambda = 9$

and $\overrightarrow{PQ}.\mathbf{n} = 0 \Rightarrow 17\mu - 11\lambda = 15.$

These equations are satisfied by $\mu = 9/8$, $\lambda = 3/8$

$$\Rightarrow \overrightarrow{PQ} = \tfrac{1}{2}\mathbf{i} - \tfrac{1}{2}\mathbf{k} \quad \text{and} \quad h = |\overrightarrow{PQ}| = 1/\sqrt{2}.$$

Equations of a plane

First consider the plane which passes through the origin and is parallel to the vectors **s** and **t**, i.e. the plane contains line segments representing **s** and **t**. If $P(\mathbf{r})$ is any point on the plane, the position vector **r** must be coplanar with **s** and **t**. Hence, by the results of §2.1, **r** may be expressed as a linear combination of **s** and **t**; that is,

$$\mathbf{r} = \lambda\mathbf{s} + \mu\mathbf{t}, \tag{4.2}$$

where λ and μ are scalars. Each point P on the plane has a position vector **r** given by equation (4.2) for some values of λ and μ, and also each pair of values of λ and μ gives a point (with position vector $\lambda\mathbf{s} + \mu\mathbf{t}$) lying on the plane. Consequently, when λ and μ are considered as variable scalar parameters, equation (4.2) is said to be an *equation of the plane* (sometimes called the vector parametric equation of the plane).

To find the equation of a plane which passes through the point $A(\mathbf{a})$ and is parallel to the vectors **s** and **t**, we note that, if $P(\mathbf{r})$ is any point on the plane, the displacement $\overrightarrow{AP}$ is coplanar with **s** and **t** (see Fig. 4.3). Thus there are scalars λ and μ such that

$$\overrightarrow{AP} = \lambda\mathbf{s} + \mu\mathbf{t}.$$

However, $\overrightarrow{AP} = \mathbf{r} - \mathbf{a}$, and so the vector parametric equation of the plane is

$$\mathbf{r} = \mathbf{a} + \lambda\mathbf{s} + \mu\mathbf{t}.$$

To find the equation of the plane through the three points $A(\mathbf{a})$, $B(\mathbf{b})$ and $C(\mathbf{c})$, we note that

$$\overrightarrow{AB} = \mathbf{b} - \mathbf{a} \quad \text{and} \quad \overrightarrow{AC} = \mathbf{c} - \mathbf{a}$$

are parallel to the plane (see Fig. 4.4). Thus the required plane passes through the point $A(\mathbf{a})$ and is parallel to $\mathbf{b} - \mathbf{a}$ and $\mathbf{c} - \mathbf{a}$. Its vector parametric equation is therefore

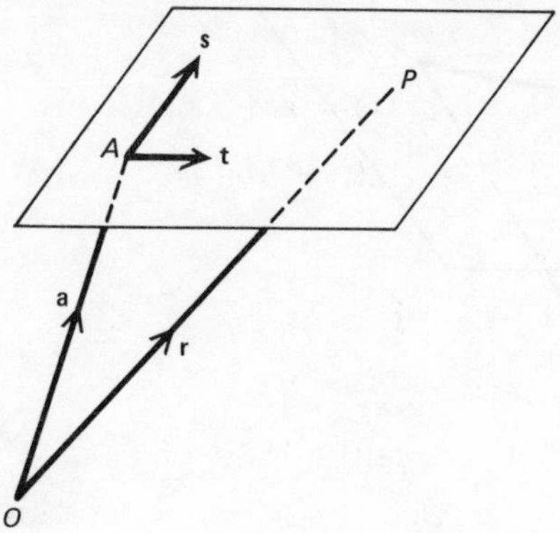

Fig. 4.3

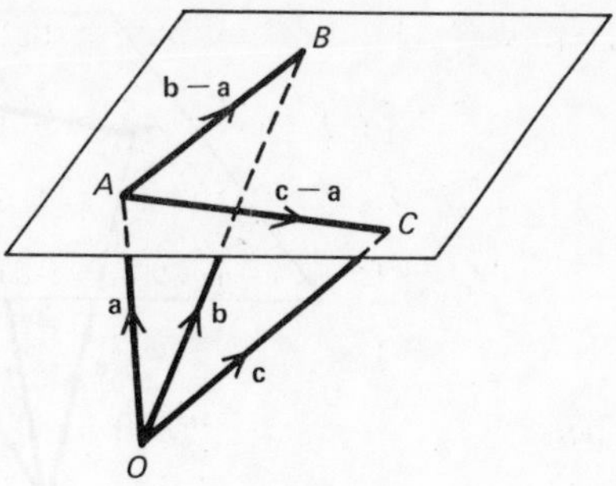

Fig. 4.4

$$\mathbf{r} = \mathbf{a} + \lambda(\mathbf{b} - \mathbf{a}) + \mu(\mathbf{c} - \mathbf{a})$$
$$= (1 - \lambda - \mu)\mathbf{a} + \lambda\mathbf{b} + \mu\mathbf{c}.$$

This equation can also be written, more symmetrically, as

$$\mathbf{r} = k\mathbf{a} + l\mathbf{b} + m\mathbf{c},$$

where $k + l + m = 1$.

Example 4 In example 2 (p. 34) we found that the lines

$$\mathbf{r} = 7\mathbf{i} - 3\mathbf{j} + 3\mathbf{k} + \lambda(3\mathbf{i} - 2\mathbf{j} + \mathbf{k})$$

and
$$\mathbf{r} = 7\mathbf{i} - 2\mathbf{j} + 4\mathbf{k} + \mu(-2\mathbf{i} + \mathbf{j} - \mathbf{k})$$

intersect at the point $A(\mathbf{i} + \mathbf{j} + \mathbf{k})$. In this case the two lines define a plane which contains them both, i.e., the plane through A and parallel to both $(3\mathbf{i} - 2\mathbf{j} + \mathbf{k})$ and $(-2\mathbf{i} + \mathbf{j} - \mathbf{k})$. The vector parametric equation of this plane is

$$\mathbf{r} = (\mathbf{i} + \mathbf{j} + \mathbf{k}) + \lambda(3\mathbf{i} - 2\mathbf{j} + \mathbf{k}) + \mu(-2\mathbf{i} + \mathbf{j} - \mathbf{k}).$$

Note that, when two lines are skew, there is no plane which contains both the lines. Such a plane exists only when the lines intersect.

Another form of the equation of a plane

Now consider the plane passing through the point $A(\mathbf{a})$ and perpendicular to the vector $\mathbf{n}$ (see Fig. 4.5).

If $P(\mathbf{r})$ lies on the plane, then $\overrightarrow{AP} = \mathbf{r} - \mathbf{a}$ is parallel to the plane and hence perpendicular to $\mathbf{n}$, so that

$$(\mathbf{r} - \mathbf{a}) . \mathbf{n} = 0. \tag{4.3}$$

It is clear from geometry that equation (4.3) is satisfied by all points $P(\mathbf{r})$

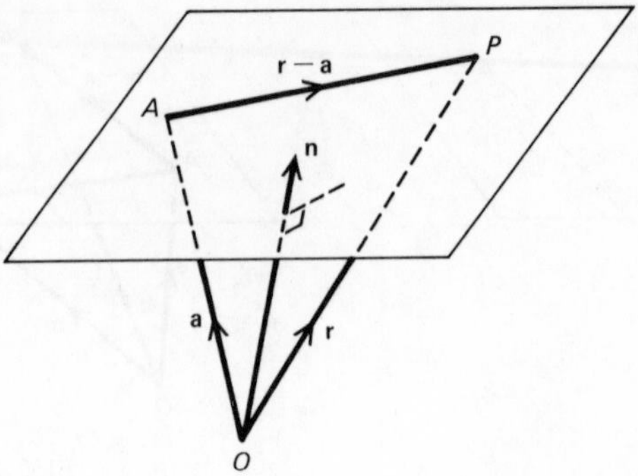

Fig. 4.5

on the plane and by no points not on the plane. We say that it is the vector equation of the plane through $A(\mathbf{a})$ perpendicular (or normal) to $\mathbf{n}$.

To find the equivalent cartesian equation of the plane, we put $\mathbf{r} = x\mathbf{i} + y\mathbf{j} + z\mathbf{k}$, $\mathbf{a} = a_1\mathbf{i} + a_2\mathbf{j} + a_3\mathbf{k}$ and $\mathbf{n} = n_1\mathbf{i} + n_2\mathbf{j} + n_3\mathbf{k}$. Then expanding the scalar product in equation (4.3) gives

$$n_1(x - a_1) + n_2(y - a_2) + n_3(z - a_3) = 0. \tag{4.4}$$

Consequently, from equation (4.4), we see that the equation

$$ax + by + cz = d \tag{4.5}$$

represents a plane, whose unit normal (i.e., a unit vector perpendicular to the plane) is the vector

$$\frac{a\mathbf{i} + b\mathbf{j} + c\mathbf{k}}{\sqrt{(a^2 + b^2 + c^2)}}.$$

Distance of a point from a plane

Using the equation of the plane in the form of equation (4.3), we see from Fig. 4.6 that the projection of $\overrightarrow{OA} = \mathbf{a}$ on $\mathbf{n}$ is equal to the perpendicular distance of the origin O from the plane. Denoting this distance by p, we have

$$p = |\mathbf{a}\,.\,\mathbf{n}|/|\mathbf{n}|.$$

Consequently, for the plane of equation (4.5), we have

$$p = |d|/\sqrt{(a^2 + b^2 + c^2)}. \tag{4.6}$$

From Fig. 4.6, it is clear that p_1, the distance of a point $P_1(\mathbf{r}_1)$ from the plane, is equal to the projection of $\overrightarrow{AP_1} = \mathbf{r}_1 - \mathbf{a}$ on $\mathbf{n}$. Thus

$$p_1 = |(\mathbf{r}_1 - \mathbf{a})\,.\,\mathbf{n}|/|\mathbf{n}|. \tag{4.7}$$

We then deduce that for the plane of equation (4.5)

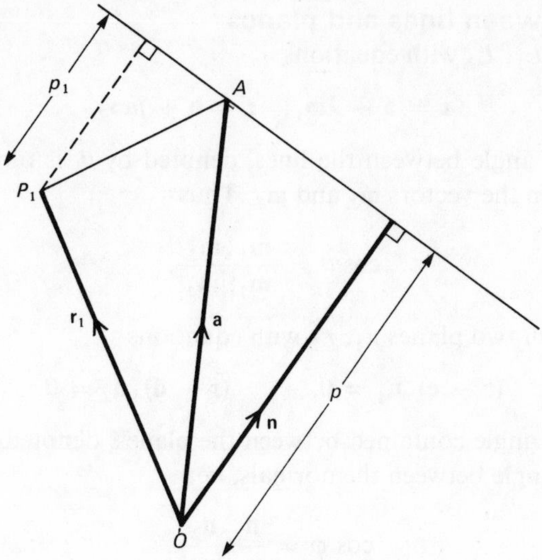

Fig. 4.6

$$p_1 = |ax_1 + by_1 + cz_1 - d|/\sqrt{(a^2 + b^2 + c^2)}, \qquad (4.8)$$

where (x_1, y_1, z_1) are the cartesian coordinates of P_1.

Example 5 Find a unit normal to the plane $6x - 3y + 2z = 5$ and the distance of the origin from this plane.

The vector $\mathbf{n} = 6\mathbf{i} - 3\mathbf{j} + 2\mathbf{k}$ is normal to the plane, and, since $6^2 + (-3)^2 + 2^2 = 7^2$, we deduce that a unit normal to the plane has direction cosines 6/7, $-3/7$, 2/7. Also, using (4.6), the origin is a distance 5/7 from the plane.

Example 6 Find a cartesian equation of the plane through the point (2, -1, -4) perpendicular to the vector $6\mathbf{i} + 2\mathbf{j} - 9\mathbf{k}$. Find also the distance of the point (4, 2, 3) from the plane.

The vector equation of the plane is

$$(\mathbf{r} - 2\mathbf{i} + \mathbf{j} + 4\mathbf{k}).(6\mathbf{i} + 2\mathbf{j} - 9\mathbf{k}) = 0$$

and putting $\mathbf{r} = x\mathbf{i} + y\mathbf{j} + z\mathbf{k}$

$$\Rightarrow 6x + 2y - 9z = 46$$

as a cartesian equation of the plane.

Then, from equation (4.8), the distance of the point (4, 2, 3) from the plane is given by p_1, where

$$p_1 = |24 + 4 - 27 - 46|/\sqrt{(36 + 4 + 81)}$$
$$= 45/11.$$

The angle between lines and planes

Given two lines L_1, L_2 with equations

$$\mathbf{r} = \mathbf{a} + \lambda\mathbf{m}_1, \quad \mathbf{r} = \mathbf{b} + \mu\mathbf{m}_2$$

respectively, the angle between the lines, denoted by θ, is obviously equal to the angle between the vectors $\mathbf{m}_1$ and $\mathbf{m}_2$. Thus

$$\cos\theta = \frac{\mathbf{m}_1 . \mathbf{m}_2}{|\mathbf{m}_1||\mathbf{m}_2|}.$$

Similarly, given two planes π_1, π_2 with equations

$$(\mathbf{r} - \mathbf{c}).\mathbf{n}_1 = 0, \qquad (\mathbf{r} - \mathbf{d}).\mathbf{n}_2 = 0$$

respectively, the angle contained between the planes, denoted by ϕ, is defined to be the *acute* angle between the normals, i.e.,

$$\cos\phi = \frac{|\mathbf{n}_1 . \mathbf{n}_2|}{|\mathbf{n}_1||\mathbf{n}_2|}.$$

Example 7 Find the angle between the two lines

$$\frac{x-1}{2} = \frac{y+2}{-9} = \frac{z+1}{6}, \qquad \frac{x+3}{12} = \frac{y}{3} = \frac{z-2}{4}.$$

The lines are parallel to the vectors

$$\mathbf{m}_1 = 2\mathbf{i} - 9\mathbf{j} + 6\mathbf{k}, \qquad \mathbf{m}_2 = 12\mathbf{i} + 3\mathbf{j} + 4\mathbf{k}$$

respectively, and the required angle θ is contained between these vectors. Therefore,

$$\cos\theta = \frac{|\mathbf{m}_1 . \mathbf{m}_2|}{|\mathbf{m}_1||\mathbf{m}_2|} = \frac{21}{11.13} = \frac{21}{143}$$

$$\Rightarrow \theta \approx 1{\cdot}42 \text{ radians} \approx 81{\cdot}6^\circ.$$

4.3 Elementary applications to kinematics

Velocity, acceleration and relative vectors

We first distinguish between the terms *speed* and *velocity*. The speed of a particle is a positive scalar quantity and gives the rate of change with time of the distance it has moved. Thus it is correct to say that a man has a speed of 5 km h^{-1}. On the other hand, velocity is a vector and is specified by a speed *and* a corresponding direction. Thus we may say that a man has a velocity of 5 km h^{-1} on a bearing of 60°. Also, whilst speed is clearly a scalar quantity, experiments show that velocities combine according to the triangle law of addition, illustrating that velocity is a vector quantity.

The position vectors **a** and **b** of two points A and B are normally given relative to a fixed origin O, so that $\mathbf{a} = \overrightarrow{OA}$ and $\mathbf{b} = \overrightarrow{OB}$. However, if we were to take the point B as a new origin, the resulting position vector of A would be $\overrightarrow{BA}$. Thus, given $A(\mathbf{a})$ and $B(\mathbf{b})$, we say that $\overrightarrow{BA} = \mathbf{a} - \mathbf{b}$ is the position vector of A relative to B.

Similarly, if $\mathbf{v}_A$ and $\mathbf{v}_B$ are the velocities of points A and B, respectively, then the velocity of A relative to B is $\mathbf{v}_A - \mathbf{v}_B$. Thus, if A and B are two moving points, the vector $\mathbf{v}_A - \mathbf{v}_B$ gives the apparent velocity of A when viewed from the (moving) point B; this is also called the *relative velocity* of A with respect to B, or the velocity of A relative to B.

Also, if $\mathbf{f}_A$ and $\mathbf{f}_B$ are the vector accelerations of points A and B, we define $\mathbf{f}_A - \mathbf{f}_B$ to be the acceleration of A relative to B.

Example 8 A man walks in a straight line from the point $A(\mathbf{a})$ to the point $B(\mathbf{b})$ with a constant speed v. To a second man, also walking at constant speed in a straight line, the first man appears to have a velocity $\mathbf{V}$. Find the velocity of the second man.

Let $\mathbf{v}_1$ and $\mathbf{v}_2$ be the velocities of the first and second man respectively. Then, as $\mathbf{V}$ is the velocity of the first man relative to the second man, we have

$$\mathbf{V} = \mathbf{v}_1 - \mathbf{v}_2 \Rightarrow \mathbf{v}_2 = \mathbf{v}_1 - \mathbf{V}.$$

To find $\mathbf{v}_1$, we note that the first man walks in the direction $\overrightarrow{AB} = \mathbf{b} - \mathbf{a}$. Thus, as $(\mathbf{b} - \mathbf{a})/|\mathbf{b} - \mathbf{a}|$ is a unit vector in this direction, the velocity of the first man is

$$\mathbf{v}_1 = v(\mathbf{b} - \mathbf{a})/|\mathbf{b} - \mathbf{a}| \Rightarrow \mathbf{v}_2 = \frac{v(\mathbf{b} - \mathbf{a})}{|\mathbf{b} - \mathbf{a}|} - \mathbf{V}.$$

When a point moves in a straight line with a constant speed u, the distance moved in time t is equal to ut. Now consider a point P moving for a time t with constant velocity $\mathbf{v} = v_1\mathbf{i} + v_2\mathbf{j} + v_3\mathbf{k}$. The velocity $\mathbf{v}$ has components v_1, v_2, v_3 along the three coordinate axes and so P will move distances of $v_1 t$, $v_2 t$, $v_3 t$ parallel to the x-, y-, z-axes respectively, in time t. Thus it has undergone a displacement equal to

$$v_1 t\mathbf{i} + v_2 t\mathbf{j} + v_3 t\mathbf{k} = \mathbf{v}t.$$

Hence if time is measured from the instant when the point P is at $A(\mathbf{a})$ (see Fig. 4.7), its position vector after time t is given by $\mathbf{r}$, where

$$\mathbf{r} = \mathbf{a} + \mathbf{v}t.$$

Of course, these results only apply when $\mathbf{v}$ is constant.

We are now able to solve problems of the following type. 'Two men P and Q are initially at points $A(\mathbf{a})$ and $B(\mathbf{b})$, respectively. Given that P moves with constant velocity $\mathbf{u}$ and Q with constant velocity $\mathbf{v}$, find the distance between P and Q at any subsequent time t'.

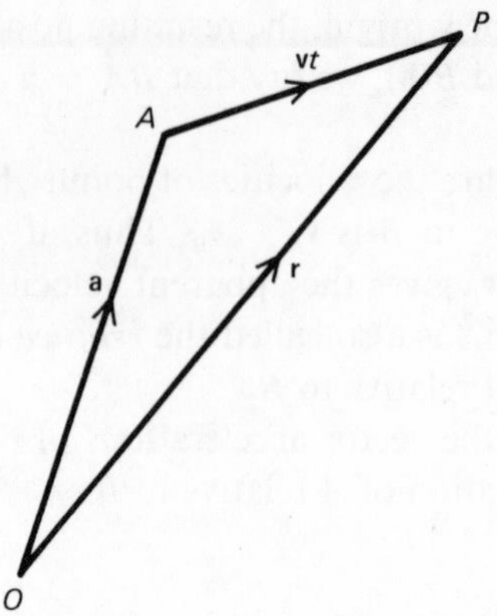

Fig. 4.7

At time t the position vectors of P and Q relative to the origin O are given by $\mathbf{p}$, $\mathbf{q}$, where

$$\mathbf{p} = \mathbf{a} + \mathbf{u}t, \quad \mathbf{q} = \mathbf{b} + \mathbf{v}t.$$

Now consider the motion relative to Q.

The position vector of P relative to Q is $\mathbf{d} = \overrightarrow{QP}$ so that

$$\mathbf{d} = \mathbf{p} - \mathbf{q} = (\mathbf{a} - \mathbf{b}) + (\mathbf{u} - \mathbf{v})t.$$

The distance between P and Q at any time is $d = |\mathbf{d}|$, which is given by

$$d^2 = \mathbf{d}.\mathbf{d} = (\mathbf{a} - \mathbf{b}).(\mathbf{a} - \mathbf{b}) + 2(\mathbf{a} - \mathbf{b}).(\mathbf{u} - \mathbf{v})t + (\mathbf{u} - \mathbf{v}).(\mathbf{u} - \mathbf{v})t^2$$

$$\Rightarrow d^2 = |\mathbf{a} - \mathbf{b}|^2 + 2t(\mathbf{a} - \mathbf{b}).(\mathbf{u} - \mathbf{v}) + t^2|\mathbf{u} - \mathbf{v}|^2.$$

Example 9 A man and a woman are standing at points A and B, respectively, where B is 15 m due east of A. The man starts walking due east at a speed of $4\ \mathrm{m\,s^{-1}}$ and at the same time the woman starts walking due north at $3\ \mathrm{m\,s^{-1}}$. Find the smallest distance between them in the subsequent motion.

Take cartesian axes with A as origin, $\mathbf{i}$ due east and $\mathbf{j}$ due north. Then the initial position vectors of the man and woman are $\mathbf{0}$ and $15\mathbf{i}$ m, respectively. As the man moves with velocity $4\mathbf{i}\ \mathrm{m\,s^{-1}}$ and the woman with velocity $3\mathbf{j}\ \mathrm{m\,s^{-1}}$, their position vectors after time t s are $\mathbf{m}$ and $\mathbf{w}$ respectively, where

$$\mathbf{m} = 4t\mathbf{i}\ \mathrm{m}, \qquad \mathbf{w} = (15\mathbf{i} + 3t\mathbf{j})\ \mathrm{m}.$$

Considering the motion of the woman relative to the man, the position vector of the woman relative to the man is $\mathbf{d}$, where

$$\mathbf{d} = \mathbf{w} - \mathbf{m} = [(15 - 4t)\mathbf{i} + 3t\mathbf{j}]\ \mathrm{m}.$$

Thus at time $\mathbf{t}$ their distance of separation $d = |\mathbf{d}|$ is given by

$$\begin{aligned} d^2 &= (\mathbf{w} - \mathbf{m}).(\mathbf{w} - \mathbf{m}) \\ &= \{(15 - 4t)^2 + 9t^2\}\ \mathrm{m}^2 \\ &= \{25t^2 - 120t + 225\}\ \mathrm{m}^2. \end{aligned}$$

To find the minimum distance, we complete the square to get

$$d^2 = \{(5t - 12)^2 + 81\}\ \mathrm{m}^2.$$

Thus the minimum value of d occurs $2\frac{2}{5}$ s after they start walking and is 9 m.

4.4 Elementary physical applications

Concurrent forces

A *force* has magnitude, measured in newtons (N), and direction, and the effect of two concurrent forces is the same as a single resultant force acting through the point of concurrency. Experiments show that the resultant is given by the vector sum of the two forces (see §1.1) and this enables us to conclude (see chapter 1) that force is a vector quantity.

Assume that a force $\mathbf{F}$ acts on a rigid body at a point $A(\mathbf{a})$. The effect, or movement of the body, produced by the force is equal to the effect produced when $\mathbf{F}$ acts at any other point on the line

$$\mathbf{r} = \mathbf{a} + \lambda\mathbf{f}, \tag{4.9}$$

where $\mathbf{f}$ is any vector parallel to $\mathbf{F}$. The force is thus said to have a *line of action* given by (4.9). If $\mathbf{F}$ acts at a point not on the line of (4.9), its effect on the body may, and generally will, be altered. Because of this, some people call force a *line vector* in order to distinguish it from free vectors and position vectors.

For the present, we confine our attention to forces whose lines of action are concurrent at a point C, so all the forces may be considered to be acting at C. This situation occurs, for example, when the forces act on a particle at C. In such a case, we can see that n concurrent forces $\mathbf{F}_1, \mathbf{F}_2, \ldots, \mathbf{F}_n$ are dynamically equivalent to a single resultant force $\mathbf{R}$, acting at C, where

$$\mathbf{R} = \mathbf{F}_1 + \mathbf{F}_2 + \ldots + \mathbf{F}_n.$$

When $\mathbf{R} = \mathbf{0}$ the forces are said to be in equilibrium.

Of particular (and historical) interest is the case when three concurrent forces are in equilibrium, so that

$$\mathbf{F}_1 + \mathbf{F}_2 + \mathbf{F}_3 = \mathbf{0}.$$

This means that the three forces are coplanar (see §2.1) and that they may be represented in magnitude and direction by the sides of a triangle taken in order (see Fig. 4.8). Applying the sine rule gives

$$\frac{|\mathbf{F}_1|}{\sin\theta_1} = \frac{|\mathbf{F}_2|}{\sin\theta_2} = \frac{|\mathbf{F}_3|}{\sin\theta_3}, \tag{4.10}$$

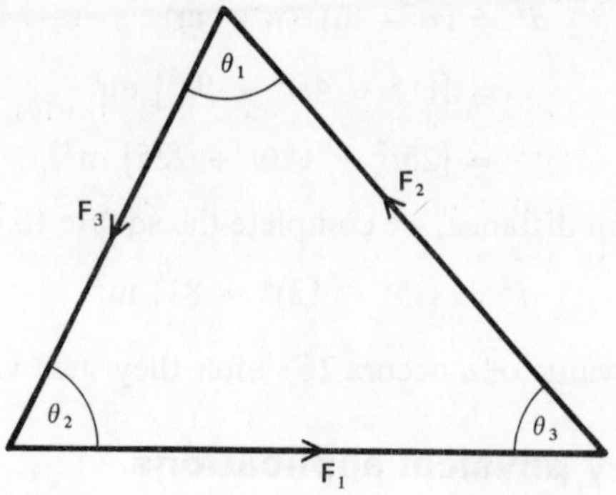

Fig. 4.8

where θ_1, θ_2, and θ_3 are the angles between the pairs of forces ($\mathbf{F}_2$, $\mathbf{F}_3$), ($\mathbf{F}_3$, $\mathbf{F}_1$) and ($\mathbf{F}_1$, $\mathbf{F}_2$), respectively. We thus have *Lami's theorem*: 'If three concurrent forces are in equilibrium they are coplanar, and the magnitude of each force is proportional to the sine of the angle between the other two'.

Work

Work is done when a force moves its point of application. The work done by a constant force is defined to be the product of the magnitude of the force and the distance moved by its point of application in the direction of the force.

So when the point of application of a constant force $\mathbf{F}$ is moved from A to B (see Fig. 4.9), we have

$$\text{work done} = |\mathbf{F}|\,|\mathbf{d}| \cos\theta = \mathbf{F}.\mathbf{d},$$

where $\overrightarrow{AB} = \mathbf{d}$ and θ is the angle between $\mathbf{d}$ and $\mathbf{F}$.

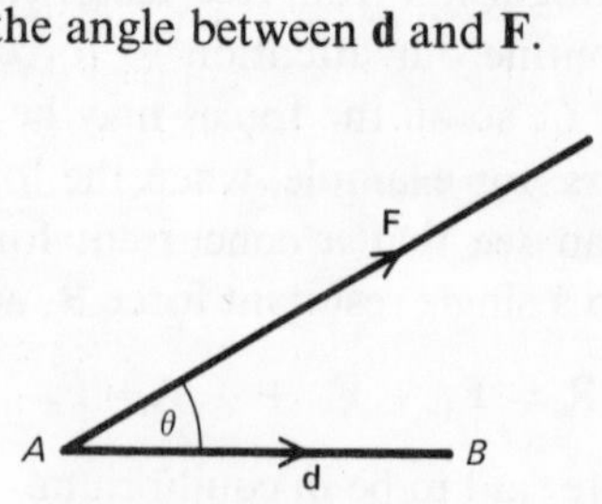

Fig. 4.9

Clearly, when $\mathbf{d}$ and $\mathbf{F}$ are perpendicular, the work done is zero. If the constant concurrent forces $\mathbf{F}_1, \mathbf{F}_2, \ldots, \mathbf{F}_n$ have their common point of application displaced by $\mathbf{d}$, the total work W done by the forces is given by

$$\begin{aligned} W &= \mathbf{F}_1.\mathbf{d} + \mathbf{F}_2.\mathbf{d} + \ldots + \mathbf{F}_n.\mathbf{d} \\ &= (\mathbf{F}_1 + \mathbf{F}_2 + \ldots + \mathbf{F}_n).\mathbf{d} \\ &= \mathbf{R}.\mathbf{d}, \end{aligned}$$

and this is the same as if the system of forces was replaced by its resultant **R**.

Example 10 Forces acting on a particle have magnitudes 4 N, 3 N, 7 N and act in the directions of the vectors $2\mathbf{i} + 9\mathbf{j} + 6\mathbf{k}$, $-6\mathbf{i} + 2\mathbf{j} + 9\mathbf{k}$, $\mathbf{j}$, respectively. The forces remain constant whilst the particle is displaced from the point $A[(3\mathbf{i} + 2\mathbf{j} - \mathbf{k})\text{ m}]$ to the point $B[(\mathbf{i} + 3\mathbf{j})\text{ m}]$. Find the work done by the forces.

The forces are $\frac{4}{11}(2\mathbf{i} + 9\mathbf{j} + 6\mathbf{k})$ N, $\frac{3}{11}(-6\mathbf{i} + 2\mathbf{j} + 9\mathbf{k})$ N, $7\mathbf{j}$ N and their resultant is therefore $\frac{1}{11}(-10\mathbf{i} + 119\mathbf{j} + 51\mathbf{k})$ N. The displacement is $(-2\mathbf{i} + \mathbf{j} + \mathbf{k})$ m, and the work done is W, where

$$W = \tfrac{1}{11}(-10\mathbf{i} + 119\mathbf{j} + 51\mathbf{k}).(-2\mathbf{i} + \mathbf{j} + \mathbf{k})\text{ N m}$$

$$= 190/11\text{ N m} = 190/11\text{ J, as } 1\text{ N m} = 1\text{ J}.$$

Exercise 4

1 The points P and Q have position vectors $\mathbf{p}$, $\mathbf{q}$ respectively. Find a vector equation of the line PQ. Given that R, S are the points with position vectors $\mathbf{p}/2$, $3\mathbf{q}/2$, find the position vector of the point of intersection of the lines PQ and RS.

2 Points A, B, C, D in a plane have position vectors $\mathbf{a} = 6\mathbf{i} + 8\mathbf{j}$, $\mathbf{b} = \frac{3}{2}\mathbf{a}$, $\mathbf{c} = 6\mathbf{i} + 3\mathbf{j}$, $\mathbf{d} = \frac{5}{3}\mathbf{c}$ respectively. Write down vector equations of the lines AD and BC and find the position vector of their point of intersection. (L)

3 The position vectors $\overrightarrow{OA}$, $\overrightarrow{OB}$ of two points A, B, not collinear with O, in the plane of the cartesian axes Oxy are $\mathbf{a}$, $\mathbf{b}$ respectively. Show that, if T is the point of intersection of AB and $A'B'$ (produced if necessary), where $\overrightarrow{OA'} = 3\mathbf{a}$, $\overrightarrow{OB'} = 2\mathbf{b}$, then $\overrightarrow{OT} = -3\mathbf{a} + 4\mathbf{b}$.

4 The position vectors of the points A, B, C are $\mathbf{a}$, $\mathbf{b}$, $\mathbf{c}$ respectively. The point P is on BC such that $BP : PC = 2 : 3$; the point Q is on CA such that $CQ : QA = 1 : 4$. Find the position vector of the point of intersection of the lines AP and BQ.

5 In the triangle ABC, O is the circumcentre and H is the orthocentre. The position vectors of A, B, C with respect to the circumcentre O as origin are $\mathbf{a}$, $\mathbf{b}$, $\mathbf{c}$ respectively. Show that an equation of the straight line through A and H is $\mathbf{r} = \mathbf{a} + t(\mathbf{b} + \mathbf{c})$, and hence find the position vector of H. (L)

6 Four coplanar points A, B, A' and B' have respective position vectors $\mathbf{a}$, $\mathbf{b}$, $3\mathbf{a}$ and $5\mathbf{b}$, where $\mathbf{a}$ and $\mathbf{b}$ are non-zero and not parallel. Find the position vector of the point of intersection of AB and $A'B'$ in terms of $\mathbf{a}$ and $\mathbf{b}$.

7 Given that $\mathbf{a}$ and $\mathbf{b}$ are unit vectors and $\mathbf{a} \neq \pm\mathbf{b}$, show that the lines L and M intersect, where L and M are given by

$$L : \mathbf{r} = \mathbf{a} + 3\mathbf{b} + s\mathbf{a}$$

$$M : \mathbf{r} = 3\mathbf{a} - 4\mathbf{b} + t\mathbf{b},$$

s and t being parameters, and find the position vector of their point of intersection, K.

Find

(a) an equation of the straight line through K perpendicular to L and M,

(b) an equation of the plane containing L and M.

8 State a vector equation of the line passing through the points A and B whose position vectors are $\mathbf{i} - \mathbf{j} + 3\mathbf{k}$ and $\mathbf{i} + 2\mathbf{j} + 2\mathbf{k}$, respectively. Determine the position vector of the point C which divides the line-segment AB internally such that $AC = 2CB$. (L)

9 A plane passes through A (2, 2, 1) and is perpendicular to the line joining the origin to A. Write a vector equation of the plane in the form $\mathbf{r}.\mathbf{n} = p$, where $\mathbf{n}$ is a unit vector. (L)

10 Find the position vector of the point P at which the line L_1, with equation

$$\mathbf{r} = (3\mathbf{i} + 5\mathbf{j} + \mathbf{k}) + \lambda(2\mathbf{i} + 6\mathbf{j} - \mathbf{k}),$$

meets the plane α, with equation

$$\mathbf{r}.(2\mathbf{i} - 3\mathbf{j} + \mathbf{k}) = 7.$$

Find also vector equations for

(a) the line L_2 which passes through P and the point with position vector $(4\mathbf{i} + 7\mathbf{j})$,

(b) the plane β which contains L_1 and L_2,

(c) the line of intersection of the planes α and β. (L)

11 Find a vector equation for the plane Π passing through the points A, B, C with position vectors $(\mathbf{i} - \mathbf{j} + 2\mathbf{k})$, $(2\mathbf{i} + \mathbf{j} + \mathbf{k})$, $(3\mathbf{i} - 2\mathbf{j} + 2\mathbf{k})$ respectively. Find also the distance from the plane Π of the point D with position vector $(3\mathbf{i} + \mathbf{j} + \mathbf{k})$.

12 Show that the lines

$$\mathbf{r} = (-2\mathbf{i} + 5\mathbf{j} - 11\mathbf{k}) + s(3\mathbf{i} + \mathbf{j} + 3\mathbf{k}),$$
$$\mathbf{r} = (8\mathbf{i} + 9\mathbf{j}) + t(4\mathbf{i} + 2\mathbf{j} + 5\mathbf{k})$$

intersect and find the position vector of their common point, P.

Show also that the plane α containing these lines is parallel to the plane β with equation

$$\mathbf{r}.(\mathbf{i} + 3\mathbf{j} - 2\mathbf{k}) = 7.$$

Find

(a) the position vector of N, the foot of the perpendicular from P to the plane β,

(b) the distance between the planes α and β. (L)

13 The point P lies on the line which is parallel to the vector $(2\mathbf{i} + \mathbf{j} - \mathbf{k})$ and which passes through the point with position vector $(\mathbf{i} + \mathbf{j} + 2\mathbf{k})$. The point Q lies on another line which is parallel to the vector $(\mathbf{i} + \mathbf{j} - 2\mathbf{k})$ and which passes through the point with position vector $(\mathbf{i} + \mathbf{j} + 4\mathbf{k})$. The line PQ is perpendicular to both these lines. Find the equation of the line PQ and the coordinates of the mid-point of PQ. (L)

14 $ABCD$ is a tetrahedron; P, Q, R, S are the centroids of the faces BCD, CDA, DAB, ABC respectively. Given that A, B, C, D have position vectors $\mathbf{a}$, $\mathbf{b}$, $\mathbf{c}$, $\mathbf{d}$, find the position vector of the point K which divides AP in the ratio 3 : 1. Show that the lines BQ, CR, DS pass through this point.

Show also that K is the mid-point of the line segments joining the mid-points of opposite edges of the tetrahedron.

15 Find the point of intersection of the line through the points (2, 0, 1) and (−1, 3, 4) and the line through the points (−1, 3, 0) and (4, −2, 5). Calculate the acute angle between the two lines. (L)

16 Referred to rectangular axes Ox and Oy, the points A, B, C, have coordinates (1, 3), (−1, 7), (4, 5) respectively.

(a) Show that the vector $\overrightarrow{OA} + \overrightarrow{OB}$ lies along the axis Oy and find its magnitude. Calculate the magnitude of the angle between the vector $\overrightarrow{OB} - \overrightarrow{OA}$ and the y-axis.

(b) The velocity of a point P is represented by the vector $3\overrightarrow{AB} + 2\overrightarrow{BC}$. Find the magnitude and the direction of the velocity of P. (L)

17 At a certain instant two ships P and Q, each moving with constant speed, are situated at the points with position vectors $(18\mathbf{i} + 7\mathbf{j})$ km and $(8\mathbf{i} + 12\mathbf{j})$ km respectively, where $\mathbf{i}$ and $\mathbf{j}$ are unit vectors due east and due north, respectively. The ratio of the speed of the ship P to the speed of the ship Q is $3:2$ and the ships are approaching each other along the straight line joining P and Q. Calculate the position vector of the point at which the ships would collide if they maintained their courses and constant speeds. (L)

18 An aircraft is flying on a straight level course at a constant speed. Its position is plotted on a radar screen. At 0930 the position vector of the aircraft plotted on the screen is $(50\mathbf{i} + 20\mathbf{j})$ cm, where $\mathbf{i}$ and $\mathbf{j}$ are unit vectors representing directions east and north respectively, referred to an origin O on the radar screen. At 0933 the position vector of the aircraft on the screen is $(62\mathbf{i} + 15\mathbf{j})$ cm. If 1 cm on the screen represents 1 km in the air, calculate

(a) the position vector of the aircraft on the screen at 0945,

(b) the velocity of the aircraft, expressed as a vector.

A second aircraft is flying at the same height, with a constant velocity represented by the vector $(210\mathbf{i} + 280\mathbf{j})$ km h^{-1}.

(c) Calculate the velocity of the first aircraft relative to the second one.

19 Two particles A and B are moving in the plane of the coordinate axes Oxy with constant velocity vectors $\mathbf{v}_1$ and $\mathbf{v}_2$, respectively, where

$$\mathbf{v}_1 = (4\mathbf{i} + 2\mathbf{j})\ \text{m s}^{-1} \quad \text{and} \quad \mathbf{v}_2 = (2\mathbf{i} + 3\mathbf{j})\ \text{m s}^{-1}.$$

Find the velocity of A relative to B. At time $t = 0$, particle A is at the point whose position vector is $(-3\mathbf{i} + 6\mathbf{j})$ m. Given that A collides with B after 4 s, find the position vector of B at $t = 0$.

20 A port P has position vector $(4\mathbf{i} + 3\mathbf{j})$ km with respect to axes Ox and Oy. A ship X leaves P and steers with constant velocity vector $(3\mathbf{i} + 2\mathbf{j})$ km h^{-1}. At the same instant another ship Y sets out from a point Q which has position vector $(5\mathbf{i} - 12\mathbf{j})$ km. Write down vectors to represent

(a) the velocity of ship Y, if the velocity of X relative to Y is $(6\mathbf{i} - 2\mathbf{j})$ km h^{-1},

(b) YX at time t h after X leaves P.

Find the shortest distance between the ships and the distance XP when the ships are closest together.

21 Two particles A and B start at the same instant from the origin O. Particle A moves with constant velocity $(2\mathbf{i} + 6\mathbf{j})$ m s^{-1} referred to O, while particle B has an initial velocity $(4\mathbf{i} - 8\mathbf{j})$ m s^{-1} and a constant acceleration $(\mathbf{i} + \mathbf{j})$ m s^{-2}.

(a) Find the position of each particle after t seconds.

(b) Find the time at which the velocities of A and B are perpendicular.

(c) Given that C is the point with position vector $(17\mathbf{i} + 25\mathbf{j})$ m, find the time at which the line joining the positions of A and B passes through C.

22 Particle A moves with velocity $(\mathbf{i} + \mathbf{j})$ m s^{-1} and particle B moves with velocity $(-\mathbf{i} + 2\mathbf{j})$ m s^{-1}. Write down the velocity of B relative to A.

Particle A is at the origin at the same instant that B is at the point with position vector $(2\mathbf{i} + \mathbf{j})$ m. Find the shortest distance between A and B in the subsequent motion.

23 Two particles A and B start simultaneously from points with position vectors $(3\mathbf{i} + 4\mathbf{j})$ m and $(8\mathbf{i} + 4\mathbf{j})$ m respectively. Each moves with constant velocity, the velocity of B being $(\mathbf{i} + 4\mathbf{j})$ m s^{-1}.

(a) Given that the velocity of A is $(5\mathbf{i} + 2\mathbf{j})$ m s^{-1}, show that the least distance between A and B in the subsequent motion is $\sqrt{5}$ m.

(b) Given alternatively that the speed of A is $5\ \mathrm{m\,s^{-1}}$, find the velocity of A if the particles collide.

24 (a) A particle P moves with constant velocity $\mathbf{v}$ and, at time $t = 0$, the position vector of P referred to a fixed origin O is $\mathbf{r}_0$. Write down the vector equation of the path of P and show that P is (or was) closest to O at time

$$t = -\frac{\mathbf{v}.\mathbf{r}_0}{\mathbf{v}.\mathbf{v}}.$$

(b) At time $t = 0$ a ship S is at a distance a due north of a fixed point O and a ship T is at a distance $2a$ due east of O. Ship S steams south-east with constant speed V and ship T steams north-east with constant speed $2V$. Write down expressions for $\overrightarrow{OS}$ and $\overrightarrow{OT}$ at time t in terms of unit vectors $\mathbf{i}$, $\mathbf{j}$ directed eastwards and northwards respectively. Deduce that O, S, T are collinear at time $t = a/V$. (L)

25 Two forces $\mathbf{P}$ and $\mathbf{Q}$ act through the point A whose position vector is $(2\mathbf{i} - 3\mathbf{j})$ m. The force $\mathbf{P}$ has magnitude $2\sqrt{13}$ N and is parallel to OA. The force $\mathbf{Q}$ has magnitude $4\sqrt{13}$ N and is parallel to AD, where D is the point with position vector $4\mathbf{i}$ m. Find the resultant of $\mathbf{P}$ and $\mathbf{Q}$.

26 A particle is in equilibrium under the action of three forces $\mathbf{P}$, $\mathbf{Q}$ and $\mathbf{R}$. Given that $\mathbf{P} = (3\mathbf{i} + 5\mathbf{j})$ N, $\mathbf{Q} = (-2\mathbf{i} + 6\mathbf{j})$ N, calculate

(a) $|\mathbf{R}|$,

(b) the tangent of the acute angle made by the line of action of $\mathbf{R}$ with the positive x-axis.

27 The forces $\mathbf{P}$ and $\mathbf{Q}$ act through the points with position vectors $(\mathbf{i} + 3\mathbf{j})$ m and $-3\mathbf{j}$ m, respectively. Given that $\mathbf{P} = 3\mathbf{i}$ N and $\mathbf{Q} = 4\mathbf{j}$ N, find the position vector of the point where the line of action of the resultant of $\mathbf{P}$ and $\mathbf{Q}$ cuts the x-axis.

28 The force $\mathbf{P} = 2\mathbf{i}$ N acts through the point with position vector $5\mathbf{j}$ m. The force $\mathbf{Q} = 6\mathbf{i}$ N acts through the point with position vector $\mathbf{j}$ m. The resultant of $\mathbf{P}$ and $\mathbf{Q}$ acts through the point with position vector $a\mathbf{j}$ m. Find the value of a and the magnitude of the resultant of $\mathbf{P}$ and $\mathbf{Q}$.

29 Forces $\mathbf{P} = (3\mathbf{i} - \mathbf{j} + 6\mathbf{k})$ N and $\mathbf{Q} = (\mathbf{i} + 5\mathbf{j} + 6\mathbf{k})$ N act through points whose position vectors are $(2\mathbf{i} - \mathbf{j} + 3\mathbf{k})$ m and $(7\mathbf{i} - 4\mathbf{j} + 12\mathbf{k})$ m, respectively. Show that the lines of action of the forces intersect. Find the position vector of their point of intersection.

30 Forces $\mathbf{F}_1 = (2\mathbf{i} + 3\mathbf{j} + 6\mathbf{k})$ N and $\mathbf{F}_2 = (2\mathbf{i} + 4\mathbf{j} + 2\mathbf{k})$ N act through the point whose position vector is $(\mathbf{i} - 2\mathbf{j} + 3\mathbf{k})$ m. Determine the values of p and q if their resultant passes through the point with position vector $(p\mathbf{i} + q\mathbf{j} + 19\mathbf{k})$ m.

31 The position vectors of the vertices L, M and N of a triangle LMN are $(2\mathbf{i} + \mathbf{j})$ m, $(5\mathbf{i} + 5\mathbf{j})$ m and $(9\mathbf{i} + 25\mathbf{j})$ m respectively. Two forces of magnitude 10 N and 75 N act along LM, LN respectively. Calculate the resultant of these forces.

32 The position vectors of points P, Q are $(2\mathbf{i} - \mathbf{j} + 2\mathbf{k})$ m, $(3\mathbf{i} + 6\mathbf{j} - 2\mathbf{k})$ m respectively. Forces of 9 N and 14 N act along OP, OQ respectively. Find the magnitude of the resultant of these forces.

33 A particle P is acted on by three forces $\mathbf{F}_1$, $\mathbf{F}_2$, $\mathbf{F}_3$, of magnitudes 6 N, 4 N, 2 N respectively and in the directions of the vectors $(6\mathbf{i} + 2\mathbf{j} + 3\mathbf{k})$, $(3\mathbf{i} - 2\mathbf{j} + 6\mathbf{k})$, $(2\mathbf{i} - 3\mathbf{j} - 6\mathbf{k})$ respectively. Prove that the work done by these forces when P is displaced from a point A with position vector $(2\mathbf{i} - \mathbf{j} - 3\mathbf{k})$ m to a point B with position vector $(5\mathbf{i} - \mathbf{j} + \mathbf{k})$ m is $39\frac{3}{7}$ J.

5 Differentiation and integration of vectors

5.1 Vectors depending on a scalar

We have already met vectors which depend on the value of a scalar variable; for example, if **r** is the position vector of a point on a straight line

$$\mathbf{r} = \mathbf{a} + \lambda\mathbf{m}$$

and the value of **r** varies as the scalar λ varies. Such vectors frequently occur in geometry and in mechanics (where the scalar variable is often time t).

Typical examples are:
(i) $\mathbf{r} = a(\mathbf{i}\cos\theta + \mathbf{j}\sin\theta)$. Then if $\mathbf{r} = x\mathbf{i} + y\mathbf{j}$, $x = a\cos\theta$, $y = a\sin\theta$ and $x^2 + y^2 = a^2$. So $P(\mathbf{r})$ is a point on a circle centre O and radius a.
(ii) $\mathbf{r} = at^2\mathbf{i} + 2at\mathbf{j}$. Then $P(\mathbf{r})$ is a point on the parabola with parametric coordinates $x = at^2$, $y = 2at$.
(iii) $\mathbf{r} = \mathbf{a} + \mathbf{v}t$. From §4.3 we know that $P(\mathbf{r})$ then moves on a straight line through the point $A(\mathbf{a})$ with velocity **v**.
(iv) $\mathbf{r} = \mathbf{u}t + \frac{1}{2}\mathbf{g}t^2$. In this case $P(\mathbf{r})$ is the position of a particle moving freely under gravity.

Any vector of the form

$$\mathbf{a} = a_1(t)\mathbf{i} + a_2(t)\mathbf{j} + a_3(t)\mathbf{k},$$

where $a_1(t)$, $a_2(t)$, $a_3(t)$ are known functions of a scalar variable t, is said to be a vector which is a function of the scalar t. The value of **a** for a particular value of t is written as $\mathbf{a}(t)$.

5.2 Differentiation of a vector with respect to a scalar

Given the point $P(\mathbf{r})$, where **r** is a function of a scalar variable t, we note that, as t varies continuously, the point P moves along a continuous curve, denoted by C, in space. Note that, in general, there is no plane containing C.

Suppose P and Q have position vectors $\mathbf{r}(t)$ and $\mathbf{r}(t + \delta t)$, respectively. Let $\delta\mathbf{r}$ be the change in $\mathbf{r}(t)$ produced by the small change δt in the scalar variable t (see Fig. 5.1) so that

$$\delta\mathbf{r} = \mathbf{r}(t + \delta t) - \mathbf{r}(t)$$

$$\Rightarrow \frac{\delta\mathbf{r}}{\delta t} = \frac{\mathbf{r}(t + \delta t) - \mathbf{r}(t)}{\delta t}.$$

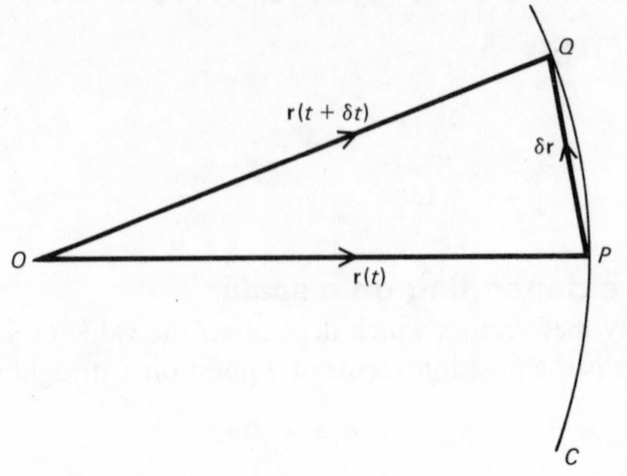

Fig. 5.1

As $\delta t \to 0$, the point Q moves along the curve C towards P and, in the limit, the chord PQ coincides in direction with the tangent to C at the point P. Hence, the limiting value of $\dfrac{\delta \mathbf{r}}{\delta t}$, as $\delta t \to 0$, is a vector whose direction is that of the tangent at P in the sense of increasing t. The limiting value of $\dfrac{\delta \mathbf{r}}{\delta t}$, denoted by $\dfrac{\mathrm{d}\mathbf{r}}{\mathrm{d}t}$, or $\dot{\mathbf{r}}$, is called the *derivative* of $\mathbf{r}$ with respect to the scalar t.

Thus
$$\frac{\mathrm{d}\mathbf{r}}{\mathrm{d}t} = \lim_{\delta t \to 0}\left(\frac{\delta \mathbf{r}}{\delta t}\right) = \lim_{\delta t \to 0}\left(\frac{\mathbf{r}(t+\delta t) - \mathbf{r}(t)}{\delta t}\right). \tag{5.1}$$

As $\dfrac{\mathrm{d}\mathbf{r}}{\mathrm{d}t}$ is also, in general, a function of the scalar t, it may be differentiated with respect to t to give the second derivative of $\mathbf{r}$, denoted by $\dfrac{\mathrm{d}^2\mathbf{r}}{\mathrm{d}t^2}$ or $\ddot{\mathbf{r}}$.

Of special interest is the case when the scalar t is time. Then $\delta \mathbf{r}$ is the displacement of P during the time interval δt and the vector $\dfrac{\delta \mathbf{r}}{\delta t}$ is the *average velocity* (where average is with respect to time) during this interval. The limiting value, as $\delta t \to 0$, then gives the *velocity* of P at time t already introduced in §4.3. Hence if $\mathbf{v}$ denotes the velocity, $\mathbf{v} = \dfrac{\mathrm{d}\mathbf{r}}{\mathrm{d}t} = \dot{\mathbf{r}}$.

Similarly, the acceleration $\mathbf{f}$ of P is given by
$$\mathbf{f} = \frac{\mathrm{d}\mathbf{v}}{\mathrm{d}t} = \frac{\mathrm{d}^2\mathbf{r}}{\mathrm{d}t^2} = \ddot{\mathbf{r}}.$$

5.3 Properties of the derivative

If $\mathbf{a}$ and $\mathbf{b}$ are two vector functions of a scalar variable t and if λ is a scalar function of t, the following results are easily established:

$$\frac{\mathrm{d}}{\mathrm{d}t}(\mathbf{a}+\mathbf{b}) = \frac{\mathrm{d}\mathbf{a}}{\mathrm{d}t} + \frac{\mathrm{d}\mathbf{b}}{\mathrm{d}t}, \tag{5.2}$$

$$\frac{\mathrm{d}}{\mathrm{d}t}(\lambda\mathbf{a}) = \mathbf{a}\frac{\mathrm{d}\lambda}{\mathrm{d}t} + \lambda\frac{\mathrm{d}\mathbf{a}}{\mathrm{d}t}, \tag{5.3}$$

$$\frac{\mathrm{d}}{\mathrm{d}t}(\mathbf{a}\,.\,\mathbf{b}) = \mathbf{a}\cdot\frac{\mathrm{d}\mathbf{b}}{\mathrm{d}t} + \frac{\mathrm{d}\mathbf{a}}{\mathrm{d}t}\cdot\mathbf{b}. \tag{5.4}$$

These results are analogous to the rules governing the differentiation of the sum and product of scalar functions of t.

A constant vector **c** is one whose magnitude and direction are constant. Thus **c** does not change with t,

and so $\quad \mathbf{c}(t+\delta t) = \mathbf{c}(t).$

Hence $\quad \dot{\mathbf{c}} = \dfrac{\mathrm{d}\mathbf{c}}{\mathrm{d}t} = \mathbf{0}.$

As the orthonormal basis vectors **i**, **j**, **k** are constant, we have

$$\frac{\mathrm{d}\mathbf{i}}{\mathrm{d}t} = \frac{\mathrm{d}\mathbf{j}}{\mathrm{d}t} = \frac{\mathrm{d}\mathbf{k}}{\mathrm{d}t} = \mathbf{0}.$$

We now deduce that if

$$\mathbf{a} = a_1(t)\mathbf{i} + a_2(t)\mathbf{j} + a_3(t)\mathbf{k},$$

then
$$\dot{\mathbf{a}} = \frac{\mathrm{d}\mathbf{a}}{\mathrm{d}t} = \frac{\mathrm{d}a_1}{\mathrm{d}t}\mathbf{i} + \frac{\mathrm{d}a_2}{\mathrm{d}t}\mathbf{j} + \frac{\mathrm{d}a_3}{\mathrm{d}t}\mathbf{k}. \tag{5.5}$$

Example 1 Given a point $P(\mathbf{r})$, where $\mathbf{r} = a(\mathbf{i}\cos\omega t + \mathbf{j}\sin\omega t)$ and a, ω are constants, find the velocity of P at time t and show that its acceleration is $-\omega^2\mathbf{r}$.

Velocity $= \dfrac{\mathrm{d}\mathbf{r}}{\mathrm{d}t} = a(-\mathbf{i}\omega\sin\omega t + \mathbf{j}\omega\cos\omega t).$

Acceleration $= \dfrac{\mathrm{d}^2\mathbf{r}}{\mathrm{d}t^2} = a(-\mathbf{i}\omega^2\cos\omega t - \mathbf{j}\omega^2\sin\omega t) = -\omega^2\mathbf{r}.$

Note that P moves with constant angular speed ω around a circle centre O and radius a. It is easily verified that $\mathbf{r}\cdot\dfrac{\mathrm{d}\mathbf{r}}{\mathrm{d}t} = 0$ and so the velocity is tangential to the circle. Note also that the acceleration is always directed towards O, the centre of the circle, and has magnitude $\omega^2 a$.

Now let **a** be any vector function of a scalar variable t whose magnitude $|\mathbf{a}|$ is constant but whose direction changes with t. Then

$$\mathbf{a}\,.\,\mathbf{a} = |\mathbf{a}|^2 = \text{constant}.$$

However,

$$\frac{\mathrm{d}}{\mathrm{d}t}(\mathbf{a}\,.\,\mathbf{a}) = \mathbf{a}\cdot\frac{\mathrm{d}\mathbf{a}}{\mathrm{d}t} + \frac{\mathrm{d}\mathbf{a}}{\mathrm{d}t}\cdot\mathbf{a} = 2\mathbf{a}\cdot\frac{\mathrm{d}\mathbf{a}}{\mathrm{d}t},$$

and, as the derivative of a constant is zero, we deduce that

$$\mathbf{a}\cdot\frac{\mathrm{d}\mathbf{a}}{\mathrm{d}t} = 0. \tag{5.6}$$

Consequently the derivative of a varying vector of constant length is always perpendicular to that vector.

Example 2 Consider the vector $\mathbf{a} = a\cos\theta\mathbf{i} + b\sin\theta\mathbf{j}$, where a and b are constants and θ is a scalar variable. As θ varies continuously the point $A(\mathbf{a})$ traces out an ellipse ($x = a\cos\theta$, $y = b\sin\theta$, $z = 0$ are the parametric equations of an ellipse in the plane $z = 0$). Find equations for the tangent and normal to the ellipse at A.

By the work in §5.2, the vector $\dfrac{\mathrm{d}\mathbf{a}}{\mathrm{d}\theta} = \mathbf{p} = -a\sin\theta\mathbf{i} + b\cos\theta\mathbf{j}$ is tangential to the ellipse at the point A.
So, if $\mathbf{r} = x\mathbf{i} + y\mathbf{j}$ is the position vector of any point on the tangent at A, there exists a number λ such that

$$\mathbf{r} = \mathbf{a} + \lambda\mathbf{p}.$$

Or, by equating components, the cartesian equation is

$$\frac{x - a\cos\theta}{-a\sin\theta} = \frac{y - b\sin\theta}{b\cos\theta}, \quad z = 0$$

$$\Rightarrow xb\cos\theta + ya\sin\theta = ab, \quad z = 0.$$

Similarly, if $\mathbf{r}$ now denotes the position vector of any point on the normal at A,

$$(\mathbf{r} - \mathbf{a})\,.\,\mathbf{p} = 0.$$

Again using components, the cartesian equation is

$$(x - a\cos\theta)(-a\sin\theta) + (y - b\sin\theta)(b\cos\theta) = 0, \quad z = 0,$$

$$\Rightarrow xa\sin\theta - yb\cos\theta = (a^2 - b^2)\sin\theta\cos\theta, \quad z = 0.$$

5.4 Integration of a vector with respect to a scalar

When a vector $\mathbf{a}$ depends on a scalar variable t, the process of finding a vector $\mathbf{A}$ such that $\dfrac{\mathrm{d}\mathbf{A}}{\mathrm{d}t} = \mathbf{a}$ is called *integration*. The vector $\mathbf{A}$ is said to be the *integral* of $\mathbf{a}$ with respect to the scalar t and we write

$$\mathbf{A} = \int \mathbf{a}\,\mathrm{d}t.$$

However, if **c** is a constant vector,

$$\frac{\mathrm{d}}{\mathrm{d}t}(\mathbf{A} + \mathbf{c}) = \frac{\mathrm{d}\mathbf{A}}{\mathrm{d}t}$$

$$\Rightarrow \mathbf{A} + \mathbf{c} = \int \mathbf{a}\,\mathrm{d}t$$

so that **A** and **A** + **c** are both integrals of **a**. The vector **A** + **c** is the most general integral of **a** and in particular problems the constant vector **c** is determined from additional data relevant to the problem.

If **a** and **b** are two vector functions of a scalar variable t, **c** is a constant vector and λ is a scalar function of t, equations (5.2) and (5.3)

$$\Rightarrow \int (\mathbf{a} + \mathbf{b})\,\mathrm{d}t = \int \mathbf{a}\,\mathrm{d}t + \int \mathbf{b}\,\mathrm{d}t \qquad (5.7)$$

and

$$\int \lambda\mathbf{c}\,\mathrm{d}t = \mathbf{c}\int \lambda\,\mathrm{d}t. \qquad (5.8)$$

Consequently, if $\mathbf{a} = a_1(t)\mathbf{i} + a_2(t)\mathbf{j} + a_3(t)\mathbf{k}$,

$$\int \mathbf{a}\,\mathrm{d}t = \mathbf{i}\int a_1(t)\,\mathrm{d}t + \mathbf{j}\int a_2(t)\,\mathrm{d}t + \mathbf{k}\int a_3(t)\,\mathrm{d}t. \qquad (5.9)$$

Example 3 Evaluate $\int (\cos 2t\mathbf{i} + \sin t\mathbf{j} + t^2\mathbf{k})\,\mathrm{d}t$.

From (5.9),

$$\begin{aligned} I &= \int (\cos 2t\mathbf{i} + \sin t\mathbf{j} + t^2\mathbf{k})\,\mathrm{d}t \\ &= \mathbf{i}\int \cos 2t\,\mathrm{d}t + \mathbf{j}\int \sin t\,\mathrm{d}t + \mathbf{k}\int t^2\,\mathrm{d}t \\ &= \mathbf{i}(\tfrac{1}{2}\sin 2t) + \mathbf{j}(-\cos t) + \mathbf{k}(\tfrac{1}{3}t^3) + \mathbf{c} \\ &= \tfrac{1}{2}\mathbf{i}\sin 2t - \mathbf{j}\cos t + \tfrac{1}{3}\mathbf{k}t^3 + \mathbf{c}, \end{aligned}$$

where **c** is a constant.

Example 4 The acceleration of a particle at time t s is equal to $(2\mathbf{i} - 6\mathbf{j}t + 12t^2\mathbf{k})\ \mathrm{m\,s^{-2}}$. Given that initially the particle is at rest at the point with position vector $(\mathbf{i} + \mathbf{j} + \mathbf{k})$ m, find its velocity and position vector at any subsequent time.

Let the particle have position vector **r**, so that its velocity is $\dot{\mathbf{r}}$ and its acceleration $\ddot{\mathbf{r}}$.

Then

$$\ddot{\mathbf{r}} = (2\mathbf{i} - 6t\mathbf{j} + 12t^2\mathbf{k})\ \mathrm{m\,s^{-2}}$$

and integrating,

$$\dot{\mathbf{r}} = (2t\mathbf{i} - 3t^2\mathbf{j} + 4t^3\mathbf{k} + \mathbf{c}_1)\ \mathrm{m\,s^{-1}}.$$

Now $\dot{\mathbf{r}} = 0$ when $t = 0$,

$$\Rightarrow \mathbf{c}_1 = \mathbf{0}.$$

$$\therefore \dot{\mathbf{r}} = (2t\mathbf{i} - 3t^2\mathbf{j} + 4t^3\mathbf{k})\ \mathrm{m\,s^{-1}}$$

and integrating again

$$\Rightarrow \mathbf{r} = (t^2\mathbf{i} - t^3\mathbf{j} + t^4\mathbf{k} + \mathbf{c}_2)\ \mathrm{m}.$$

As $\quad \mathbf{r} = (\mathbf{i} + \mathbf{j} + \mathbf{k})\ \mathrm{m} \quad$ when $t = 0$

$$\mathbf{i} + \mathbf{j} + \mathbf{k} = \mathbf{c}_2$$

and $\quad \mathbf{r} = [(1 + t^2)\mathbf{i} + (1 - t^3)\mathbf{j} + (1 + t^4)\mathbf{k}]\ \mathrm{m}.$

Exercise 5

1 At time t s, the coordinates of a moving point A are $(4t^2, 8t)$ m.
(a) Find the components of the velocity of A parallel to the axes Ox and Oy at time t s.
(b) Show that the acceleration of A is parallel to Ox.
(c) Given that the velocity of a second point B relative to A at time t s is $(13t\mathbf{i} - 6\mathbf{j})\ \mathrm{m\,s^{-1}}$, calculate the acceleration of B.

2 Two particles P and Q are moving in a horizontal plane. At time t seconds P has position vector $(t^2\mathbf{i} + 2t\mathbf{j})$ m and Q has position vector $(3t\mathbf{i} - \mathbf{j})$ m.
(a) Find the position vectors of P and Q when $t = 3$, and the distance PQ at this instant.
(b) Show that Q is moving with constant velocity and find the velocity of P when $t = 1$.
(c) Find the velocity of P relative to Q when $t = 1$.

3 Two particles P and Q are moving in a horizontal plane. At time t s, P and Q have position vectors $(\frac{3}{2}t^2 + 3t)\mathbf{i}$ m and $[\frac{1}{6}t^3\mathbf{i} - (t^2 - 5t)\mathbf{j}]$ m respectively.
(a) Find the velocities of P and Q when $t = 2$.
(b) Find the velocity of P relative to Q when $t = 2$.
(c) Find the acceleration vectors of P and Q when $t = 3$.

4 At time t, two particles A and B have position vectors $(2a \sin \omega t)\mathbf{i} + (2a \cos \omega t)\mathbf{j}$ and $2\omega t a\mathbf{i} + \omega^2 t^2 a\mathbf{j}$ respectively, where ω and a are positive constants. Find the cartesian equations of the paths followed by A and B. Find also
(a) the magnitude of the velocity of A relative to B when $t = 0$,
(b) the magnitude of the acceleration of A relative to B when $t = \pi/(2\omega)$.
Find also the values of t for which
(c) the accelerations of A and B are parallel and in the same sense,
(d) the accelerations of A and B are parallel and in opposite senses.

5 The position vector of a particle at time t is given by

$$\mathbf{r} = a(2 \cos 2\omega t)\mathbf{i} + a(2 \sin 2\omega t)\mathbf{j} + a(3\omega^2 t^2 - 1)\mathbf{k}.$$

Show that the speed of the particle increases with time but that the magnitude of the acceleration remains constant. (L)

6 The points S and T are moving anti-clockwise round a circle with centre O and radius

a, with the same constant speed $a\omega$. When T is passing through the point with position vector $a\mathbf{j}$, S is passing through the point with position vector $a\mathbf{i}$. Find, at time t later, the position vectors of S and T. Find also, in vector form, the velocity of T relative to S. (L)

7 At time t s, the position vector $\overrightarrow{OP}$ is $(2\mathbf{i} + t\mathbf{j})$ m. Write down an expression for $\tan\theta$, where θ is the angle which $\overrightarrow{OP}$ makes with the vector $\mathbf{i}$, and deduce that the angular speed of $\overrightarrow{OP}$ at time t is $2/(4 + t^2)\ \mathrm{s}^{-1}$.

8 A particle moves with constant acceleration $\mathbf{f}$. If at time $t = 0$ the particle is at the origin O and has velocity $\mathbf{V}$, find its position vector $\mathbf{r}$ at time t and show that its path lies in a plane. Show also that the path of the particle is, in general, a parabola. (L)

9 (a) At time t s, the respective position vectors $\mathbf{r}_1$ and $\mathbf{r}_2$ of two particles P_1 and P_2 are given by

$$\mathbf{r}_1 = [2t\mathbf{i} + (3t^2 - 4t)\mathbf{j} - t^3\mathbf{k}]\ \mathrm{m}$$

$$\mathbf{r}_2 = [t^3\mathbf{i} - 2t\mathbf{j} + (2t^2 - 1)\mathbf{k}]\ \mathrm{m}.$$

Find the velocity and acceleration of P_2 relative to P_1 when $t = 2$.

(b) The acceleration $\mathbf{a}$ of a particle A at time t s is given by

$$\mathbf{a} = [2\mathrm{e}^{-t}\mathbf{i} + (5\cos t)\mathbf{j} + (3\sin t)\mathbf{k}]\ \mathrm{m\,s^{-2}}.$$

Given that, at $t = 0$, A is at the point with position vector $(\mathbf{i} + \mathbf{j} + \mathbf{k})$ m and has velocity $(\mathbf{i} + 3\mathbf{j} + \mathbf{k})\ \mathrm{m\,s^{-1}}$, find the velocity and position vector of A at time t s.

10 At time $t = 0$, a particle is projected from the origin with velocity $\mathbf{u}$ and moves with constant acceleration $\mathbf{f}$. A second particle, moving with the same acceleration, is projected simultaneously from the origin with velocity $\mathbf{v}$. Show that if $\mathbf{u}\cdot\mathbf{v} < 0$ and neither $\mathbf{u}$ nor $\mathbf{v}$ is in the opposite direction to $\mathbf{f}$, there is just one instant at which the particles subtend a right angle at the origin. Explain why this result does not hold if $\mathbf{u}$ or $\mathbf{v}$ is in the opposite direction to $\mathbf{f}$.

11 A unit vector $\mathbf{a}$ rotates in the xy plane. At time t the vector makes an angle θ with the axis Ox, the positive sense of measurement of θ being the sense of rotation from Ox to Oy. Show that

$$\frac{\mathrm{d}\mathbf{a}}{\mathrm{d}t} = \mathbf{b}\frac{\mathrm{d}\theta}{\mathrm{d}t},$$

where $\mathbf{b}$ is a unit vector perpendicular to $\mathbf{a}$, and specify the sense of $\mathbf{b}$. Obtain a similar result for $\dfrac{\mathrm{d}\mathbf{b}}{\mathrm{d}t}$.

A point P moves on a circle of radius r and centre O. At time t, OP makes an angle θ with a fixed radius of the circle. Find, in terms of θ and its time derivatives, and of r, the components of the acceleration of P along and perpendicular to $\overrightarrow{OP}$.

6 The vector product

6.1 The vector product

The *vector product* (often called the cross product) of two vectors **a** and **b**, denoted by $\mathbf{a} \times \mathbf{b}$, or sometimes $\mathbf{a} \wedge \mathbf{b}$, is defined to be a vector of magnitude $|\mathbf{a}|\,|\mathbf{b}| \sin \theta$, where θ is the angle contained between **a** and **b**, with the direction of $\mathbf{a} \times \mathbf{b}$ being perpendicular to both **a** and **b** in the sense in which a right-handed corkscrew moves when rotated from **a** to **b**. Thus if **n** is a unit vector in this direction (see Fig. 6.1)

$$\mathbf{a} \times \mathbf{b} = (|\mathbf{a}|\,|\mathbf{b}| \sin \theta)\mathbf{n}. \tag{6.1}$$

The ordered set of vectors $\{\mathbf{a}, \mathbf{b}, \mathbf{n}\}$ is said to be a *right-handed set*. (This is because of the way in which the direction of **n** is defined.) As $\{\mathbf{b}, \mathbf{a}, -\mathbf{n}\}$ is also a right-handed set of vectors, it follows that

$$\mathbf{a} \times \mathbf{b} = -\mathbf{b} \times \mathbf{a} \tag{6.2}$$

and so the vector product is not commutative.

The distributive law

$$\mathbf{a} \times (\mathbf{b} + \mathbf{c}) = \mathbf{a} \times \mathbf{b} + \mathbf{a} \times \mathbf{c}$$

holds for vector products, but we do not include a proof here.

When **i**, **j**, **k** are orthonormal basis vectors in a right-handed set of cartesian axes, it follows, directly from (6.1), that

$$\mathbf{j} \times \mathbf{k} = \mathbf{i}, \qquad \mathbf{k} \times \mathbf{i} = \mathbf{j}, \qquad \mathbf{i} \times \mathbf{j} = \mathbf{k}. \tag{6.3}$$

Also, as $\mathbf{a} \times \mathbf{a} = \mathbf{0}$ for any vector, since $\sin \theta = 0$,

$$\mathbf{i} \times \mathbf{i} = \mathbf{j} \times \mathbf{j} = \mathbf{k} \times \mathbf{k} = \mathbf{0}. \tag{6.4}$$

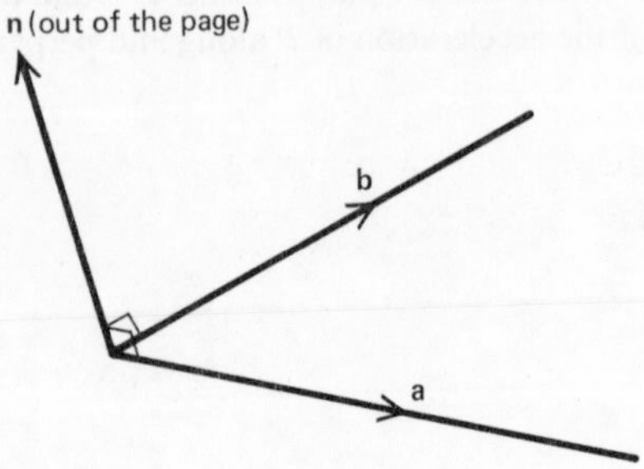

Fig. 6.1

Then if $\mathbf{a} = a_1\mathbf{i} + a_2\mathbf{j} + a_3\mathbf{k}$ and $\mathbf{b} = b_1\mathbf{i} + b_2\mathbf{j} + b_3\mathbf{k}$, it follows that

$$\mathbf{a} \times \mathbf{b} = (a_2b_3 - a_3b_2)\mathbf{i} + (a_3b_1 - a_1b_3)\mathbf{j} + (a_1b_2 - a_2b_1)\mathbf{k}. \quad (6.5)$$

Note: As the components of $\mathbf{a} \times \mathbf{b}$ are second-order determinants contained in the matrix

$$\begin{bmatrix} a_1 & a_2 & a_3 \\ b_1 & b_2 & b_3 \end{bmatrix},$$

the expression of (6.5) may be written in the symbolic form

$$\mathbf{a} \times \mathbf{b} = \begin{vmatrix} \mathbf{i} & \mathbf{j} & \mathbf{k} \\ a_1 & a_2 & a_3 \\ b_1 & b_2 & b_3 \end{vmatrix}. \quad (6.6)$$

Example 1 Given that

$$\mathbf{a} = 2\mathbf{i} - \mathbf{j} - 3\mathbf{k}, \qquad \mathbf{b} = \mathbf{i} + 2\mathbf{j} - 4\mathbf{k},$$

find $\mathbf{a} \times \mathbf{b}$.

Using (6.5) we see that

$$\begin{aligned} \mathbf{a} \times \mathbf{b} &= (2\mathbf{i} - \mathbf{j} - 3\mathbf{k}) \times (\mathbf{i} + 2\mathbf{j} - 4\mathbf{k}) \\ &= \mathbf{i}(4 + 6) + \mathbf{j}(-3 + 8) + \mathbf{k}(4 + 1) \\ &= 10\mathbf{i} + 5\mathbf{j} + 5\mathbf{k}. \end{aligned}$$

Note: Check that $\mathbf{a}.(\mathbf{a} \times \mathbf{b}) = \mathbf{b}.(\mathbf{a} \times \mathbf{b}) = 0$ and so verify that $\mathbf{a} \times \mathbf{b}$ is perpendicular to both $\mathbf{a}$ and $\mathbf{b}$ as required by the definition (6.1).

6.2 The area of a triangle

Let A and B be two points whose position vectors relative to an origin O are $\mathbf{a}$ and $\mathbf{b}$ respectively. If θ is the acute (or obtuse) angle AOB, the area of the triangle OAB is given by Δ_1, where

$$\begin{aligned} \Delta_1 &= \tfrac{1}{2}(OA)(OB)\sin\theta \\ &= \tfrac{1}{2}|\mathbf{a}|\,|\mathbf{b}|\sin\theta \\ &= \tfrac{1}{2}|\mathbf{a} \times \mathbf{b}|. \end{aligned}$$

Similarly, if C is an additional point with position vector $\mathbf{c}$ relative to O, the area of the triangle ABC is given by Δ_2, where

$$\begin{aligned} \Delta_2 &= \tfrac{1}{2}(AB)(AC)\sin BAC \\ &= \tfrac{1}{2}|(\mathbf{b} - \mathbf{a})|\,|(\mathbf{c} - \mathbf{a})|\sin BAC \\ &= \tfrac{1}{2}|(\mathbf{b} - \mathbf{a}) \times (\mathbf{c} - \mathbf{a})| \\ &= \tfrac{1}{2}|\mathbf{b} \times \mathbf{c} + \mathbf{c} \times \mathbf{a} + \mathbf{a} \times \mathbf{b}|. \end{aligned}$$

Example 2 Find the area Δ of the triangle whose vertices A, B, C have cartesian coordinates $(a, 0, 2a)$, $(-a, a, -a)$, $(3a, 2a, 0)$, respectively.

$$\overrightarrow{OA} = a(\mathbf{i} + 2\mathbf{k}), \quad \overrightarrow{OB} = a(-\mathbf{i} + \mathbf{j} - \mathbf{k}), \quad \overrightarrow{OC} = a(3\mathbf{i} + 2\mathbf{j})$$

$$\Rightarrow \overrightarrow{AB} = a(-2\mathbf{i} + \mathbf{j} - 3\mathbf{k}), \quad \overrightarrow{AC} = a(2\mathbf{i} + 2\mathbf{j} - 2\mathbf{k})$$

$$\Rightarrow \quad \overrightarrow{AB} \times \overrightarrow{AC} = a^2(4\mathbf{i} - 10\mathbf{j} - 6\mathbf{k})$$

$$\Rightarrow |\overrightarrow{AB} \times \overrightarrow{AC}| = a^2\sqrt{152} = 2a^2\sqrt{38}$$

$$\Rightarrow \quad \Delta = a^2\sqrt{38}.$$

6.3 The moment of a force

Let $A(\mathbf{a})$ be a point on the line of action of a force $\mathbf{F}$. The *moment of the force* about the origin O is defined as the vector $\mathbf{G}$, where

$$\mathbf{G} = \mathbf{a} \times \mathbf{F}.$$

In Fig. 6.2 the point N is the foot of the perpendicular from O to the line of action of the force $\mathbf{F}$. Then

$$|\mathbf{G}| = |\mathbf{a} \times \mathbf{F}| = |\mathbf{a}||\mathbf{F}| \sin \theta = |\mathbf{F}| . ON$$

which is independent of the choice of the point A on the line of action of $\mathbf{F}$, and is consistent with the definition of moment about an axis as used in old books on statics which do not use vector notation.

The moment of the force about a point $B(\mathbf{b})$ is similarly given by

$$(\mathbf{a} - \mathbf{b}) \times \mathbf{F},$$

since $(\mathbf{a} - \mathbf{b})$ is the position vector of A relative to B.

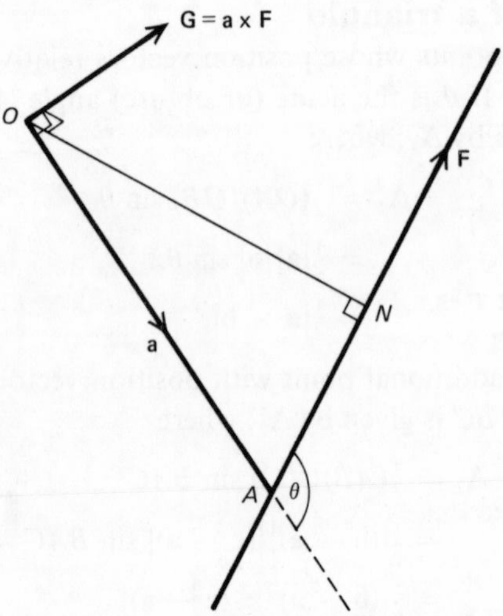

Fig. 6.2

Example 3 A force $\mathbf{F} = (2\mathbf{i} - \mathbf{j} + 3\mathbf{k})$ N acts through the point $A(\mathbf{a})$, where $\mathbf{a} = (3\mathbf{i} + \mathbf{j} - \mathbf{k})$ m. Find the moment of $\mathbf{F}$ about the point $B(\mathbf{b})$, where $\mathbf{b} = (\mathbf{i} - 2\mathbf{j} + 3\mathbf{k})$ m.

As $$\mathbf{a} - \mathbf{b} = (2\mathbf{i} + 3\mathbf{j} - 4\mathbf{k})\text{ m}$$

the moment $$\mathbf{G} = (2\mathbf{i} + 3\mathbf{j} - 4\mathbf{k}) \times (2\mathbf{i} - \mathbf{j} + 3\mathbf{k})\text{ N m}$$

$$= (5\mathbf{i} - 14\mathbf{j} - 8\mathbf{k})\text{ N m}.$$

6.4 The shortest distance between two straight lines.

Consider the two skew lines L_1 and L_2 with equations

$$\mathbf{r} = \mathbf{a} + \lambda\mathbf{m} \quad \text{and} \quad \mathbf{r} = \mathbf{b} + \mu\mathbf{n}$$

respectively. Figure 4.2 shows points P and Q on L_1 and L_2, respectively, such that the minimum distance h between the lines is given by $h = |\overrightarrow{PQ}|$. As $\overrightarrow{PQ}$ is perpendicular to both $\mathbf{m}$ and $\mathbf{n}$ and the lines are skew, the vector $\mathbf{m} \times \mathbf{n}$ is non-zero and parallel to $\overrightarrow{PQ}$. Then, as $A(\mathbf{a})$ and $B(\mathbf{b})$ lie on L_1 and L_2, respectively, the minimum distance h is equal to the projection of $\overrightarrow{AB}$ on $\mathbf{m} \times \mathbf{n}$

$$\Rightarrow h = |(\mathbf{b} - \mathbf{a}).(\mathbf{m} \times \mathbf{n})|/|\mathbf{m} \times \mathbf{n}|.$$

Compare this treatment with that in §4.2. Apply this formula to the lines in Example 3 of Chapter 4 (p. 36) and verify that it gives the same result.

Exercise 6

1 Given that $\mathbf{p} = 3\mathbf{i} - 3\mathbf{k}$ and $\mathbf{q} = \mathbf{i} + 2\mathbf{j} - 7\mathbf{k}$, calculate
(a) the scalar product $\mathbf{p}.\mathbf{q}$,
(b) the vector product $\mathbf{p} \times \mathbf{q}$. (L)

2 If $\mathbf{a}$, $\mathbf{b}$ are constant vectors, show that the equation

$$\mathbf{r} = \mathbf{a} \cos \omega t + \mathbf{b} \sin \omega t,$$

where ω is a constant, represents an ellipse. Verify that

$$\frac{d^2\mathbf{r}}{dt^2} + \omega^2\mathbf{r} = \mathbf{0} \quad \text{and} \quad \mathbf{r} \times \frac{d\mathbf{r}}{dt} = \omega\mathbf{a} \times \mathbf{b}.$$ (L)

3 The direction of a straight line through O is that of the unit vector $\mathbf{n}$. The position vector of the point P is $\mathbf{r}$. Prove that the perpendicular distance from P to the line is $|\mathbf{r} \times \mathbf{n}|$. Hence, or otherwise, find the perpendicular distance from the point (3, 1, 2) to the straight line

$$\frac{x-2}{3} = \frac{y-3}{2} = \frac{z-1}{2}.$$

4 The vectors $\mathbf{u}$ and $\mathbf{v}$ are given by

$$\mathbf{u} = 2\mathbf{i} - \mathbf{j} + 2\mathbf{k}, \qquad \mathbf{v} = p\mathbf{i} + q\mathbf{k}.$$

Given that $\mathbf{u} \times \mathbf{v} = \mathbf{i} + s\mathbf{k}$, find p, q, and s. Find also, in surd form, the cosine of the angle between $\mathbf{u}$ and $\mathbf{v}$.

5 Given that

$$\mathbf{r} = a\mathbf{i} + b\mathbf{j} + c\mathbf{k},$$

and $$\mathbf{k} \times \mathbf{r} = \mathbf{p}, \quad \mathbf{r} \times \mathbf{p} = \mathbf{k},$$

where a, b, c are constants, show that

$$a^2 + b^2 = 1 \quad \text{and} \quad c = 0.$$

6 Find the velocity $\mathbf{v}$ and acceleration $\mathbf{f}$ of a particle which moves so that, at time t, its position vector is

$$\mathbf{r} = a(\mathbf{i} \sin \omega t + \mathbf{j} \cos \omega t + \mathbf{k}\omega t),$$

where a and ω are positive constants.

Find also the times at which

$$(\mathbf{r} \times \mathbf{f}) . \mathbf{j} = 0.$$

7 The vector product of two non-zero vectors $\mathbf{p}$ and $\mathbf{q}$ is $\mathbf{p} \times \mathbf{q} = \mathbf{0}$. Show that $\mathbf{p} = \lambda \mathbf{q}$, where λ is a scalar.

Three vectors $\mathbf{a}$, $\mathbf{b}$, $\mathbf{c}$ are such that $\mathbf{b} \times \mathbf{c} = \mathbf{c} \times \mathbf{a} \neq \mathbf{0}$.

(a) Prove that $\mathbf{a} + \mathbf{b} = k\mathbf{c}$, where k is a scalar.

(b) If also $\mathbf{a} \times \mathbf{b} = \mathbf{b} \times \mathbf{c} = \mathbf{c} \times \mathbf{a} \neq \mathbf{0}$, prove that

$$\mathbf{a} + \mathbf{b} + \mathbf{c} = \mathbf{0}.$$

Give geometrical interpretations of the last two equations.

8 Show that the vector equation

$$\mathbf{r} \times \mathbf{a} = \mathbf{b},$$

where $\mathbf{a}$ and $\mathbf{b}$ are given vectors, has no solution unless $\mathbf{a} . \mathbf{b} = 0$. Solve for $\mathbf{r}$ when this condition is satisfied. (L)

9 A force of magnitude 4 N acts through the point with position vector $(4\mathbf{i} - \mathbf{j} + 7\mathbf{k})$ m in the direction of the vector $(9\mathbf{i} + 6\mathbf{j} - 2\mathbf{k})$. Find the moment of the force about the point with position vector $(\mathbf{i} - 3\mathbf{j} + 2\mathbf{k})$ m.

10 The lines of action of forces $\mathbf{F}_1$ and $\mathbf{F}_2$, where

$$\mathbf{F}_1 = (2\mathbf{i} - 3\mathbf{j} + \mathbf{k})\ \text{N} \quad \text{and} \quad \mathbf{F}_2 = (5\mathbf{i} - 4\mathbf{j} + 2\mathbf{k})\ \text{N},$$

pass through the points with position vectors $(2\mathbf{i} - \mathbf{j} + 2\mathbf{k})$ m and $(2\mathbf{i} + \mathbf{j})$ m, respectively. Calculate the total moment of these forces about the point with position vector $(\mathbf{i} + \mathbf{j} - \mathbf{k})$ m.

11 Forces $\mathbf{F}_1$ and $\mathbf{F}_2$, where

$$\mathbf{F}_1 = (\mathbf{i} + 2\mathbf{j} + 3\mathbf{k})\ \text{N} \quad \text{and} \quad \mathbf{F}_2 = (2\mathbf{i} + \mathbf{k})\ \text{N},$$

act at points with position vectors $(2\mathbf{i} + 5\mathbf{j} + b\mathbf{k})$ m and $(5\mathbf{i} + b\mathbf{j} + 2\mathbf{k})$ m, respectively. Given that these forces meet at a point, find the value of b and determine a vector equation of the line of action of the resultant of forces $\mathbf{F}_1$ and $\mathbf{F}_2$.

Show that the sum of the moments of these forces about the origin is $(12\mathbf{i} - 4\mathbf{j} - 7\mathbf{k})$ N m.

12 A plane passes through three points U, V, W whose position vectors, referred to the origin O, are $(\mathbf{i} + 3\mathbf{j} + 3\mathbf{k})$ m, $(3\mathbf{i} + \mathbf{j} + 4\mathbf{k})$ m and $(2\mathbf{i} + 4\mathbf{j} + \mathbf{k})$ m, respectively. Find, in the form $(\lambda\mathbf{i} + \mu\mathbf{j} + \nu\mathbf{k})$, a unit vector normal to this plane.

Find also a cartesian equation of the plane, and the perpendicular distance from the origin to this plane.

13 The non-collinear fixed points A, B, C have non-zero position vectors $\mathbf{a}$, $\mathbf{b}$, $\mathbf{c}$. The variable point P, in three dimensions, has position vector $\mathbf{r}$. Describe the following loci, where λ, μ are scalar parameters:

(a) $\mathbf{r} = \mathbf{a} + \lambda(\mathbf{b} - \mathbf{a})$,

(b) $\mathbf{r} = \mathbf{a} + \lambda(\mathbf{b} - \mathbf{a}) + \mu(\mathbf{c} - \mathbf{a})$,

(c) $|\mathbf{r} - \mathbf{a}| = |\mathbf{r} - \mathbf{b}|$,

(d) $|\mathbf{r} - \mathbf{a}| = |\mathbf{r} - \mathbf{b}| = |\mathbf{r} - \mathbf{c}|$,

(e) $\mathbf{r} = \lambda(\mathbf{b} - \mathbf{a}) \times (\mathbf{c} - \mathbf{a})$.

Write down the subset of the above equations which determines the centre of the circle ABC.

Answers

Exercise 1

1 $3\mathbf{b}$, $\mathbf{a}+\mathbf{b}$, $\mathbf{a}-2\mathbf{b}$
2 $\frac{1}{2}(\mathbf{b}+\mathbf{c})$, $\frac{1}{3}(\mathbf{a}+\mathbf{b}+\mathbf{c})$
3 $\frac{1}{2}(\mathbf{y}+\mathbf{z})$, $\frac{1}{3}(\mathbf{x}+\mathbf{y}+\mathbf{z})$
4 $\overrightarrow{YZ}=-\frac{1}{2}\mathbf{p}$, $\overrightarrow{ZX}=-\frac{1}{2}\mathbf{q}$, $\overrightarrow{XY}=\frac{1}{2}\mathbf{p}+\frac{1}{2}\mathbf{q}$, $\overrightarrow{RX}=-\frac{1}{2}\mathbf{p}$, $\overrightarrow{RY}=\frac{1}{2}\mathbf{q}$, $\overrightarrow{RZ}=\frac{1}{2}\mathbf{q}-\frac{1}{2}\mathbf{p}$, $\overrightarrow{RG}=\frac{1}{3}\mathbf{q}-\frac{1}{3}\mathbf{p}$
5 $\frac{1}{5}(3\mathbf{a}+2\mathbf{b})$, $\frac{1}{3}(\mathbf{a}+\mathbf{b})$
6 (a) $6(\mathbf{b}-\mathbf{a})$, $2\mathbf{a}+4\mathbf{b}$, $2(\mathbf{b}-\mathbf{a})$; (c) 12/5
7 (i) 3 : 4, (ii) $\frac{1}{2}(\mathbf{p}-\mathbf{q})$, $\frac{1}{2}(\mathbf{q}-2\mathbf{p})$, $\frac{1}{2}(\mathbf{p}-2\mathbf{q})$
8 (i) (a) 1, (b) $\frac{1}{2}$, (c) $\frac{1}{3}$; (ii) $\mathbf{b}+\mathbf{c}$, $\mathbf{c}$
10 $2\pi/3$
11 7
12 $-\mathbf{p}$, $2\mathbf{q}-\mathbf{p}$, $\mathbf{q}-2\mathbf{p}$
15 $\frac{1}{3}(\mathbf{a}+\mathbf{b}+\mathbf{c})$
16 (a) $\frac{1}{2}\mathbf{a}$, $\frac{1}{2}(\mathbf{a}+\mathbf{b})$, $\frac{1}{2}(\mathbf{b}+\mathbf{c})$, $\frac{1}{2}\mathbf{c}$; (b) $\frac{1}{2}\mathbf{b}$, $\frac{1}{2}\mathbf{b}$
17 $5\mathbf{q}=\mathbf{r}+4\mathbf{p}$, 1 : 4
18 $Y:\frac{1}{7}(4\mathbf{a}+2\mathbf{b}+\mathbf{c})$
$Z:\frac{1}{7}(\mathbf{a}+4\mathbf{b}+2\mathbf{c})$
19 $P:(s\mathbf{a}+s^2\mathbf{b}+\mathbf{c})/m$
$Q:(\mathbf{a}+s\mathbf{b}+s^2\mathbf{c})/m$
$R:(s^2\mathbf{a}+\mathbf{b}+s\mathbf{c})/m$
$m=1+s+s^2$

Exercise 2

1 (a) 17, (b) $(3\frac{1}{2}, -1)$
4 $-3\mathbf{i}-\mathbf{j}$, $-2\mathbf{i}+3\mathbf{j}$
5 3, -2
6 (a) $\overrightarrow{OA}+\overrightarrow{OC}$, (b) $\overrightarrow{OA}-\overrightarrow{OC}$, (c) $\frac{1}{2}(\overrightarrow{OA}+\overrightarrow{OC})$, (d) $(1\frac{1}{2}, -4\frac{1}{2})$, (e) $\sqrt{20}$, arc cos $(2/\sqrt{5})$
7 $\frac{1}{9}(13\mathbf{i}+18\mathbf{j})$
8 $\frac{1}{3}(2\mathbf{a}+\mathbf{b})$, $h=\frac{3}{2}$, $k=\frac{1}{2}$
9 (a) $\dfrac{1}{\sqrt{2}}(-\mathbf{i}+\mathbf{j})$, (b) $\frac{3}{2}\sqrt{29}$
10 (a) $8\mathbf{i}+15\mathbf{j}$, (b) $\frac{24}{5}(\mathbf{i}+\mathbf{j})$
11 $\mathbf{v}_3=2\mathbf{v}_1+\mathbf{v}_2$
12 3, 2
13 (i) equilateral, (ii) right-angled at A
14 $-2\mathbf{i}+2\mathbf{j}+3\mathbf{k}$
15 $\mathbf{a}=2\mathbf{b}+\mathbf{c}$, $\mathbf{b}=\frac{1}{2}\mathbf{a}-\frac{1}{2}\mathbf{c}$, $\mathbf{c}=\mathbf{a}-2\mathbf{b}$
16 $\frac{1}{9}(29\mathbf{a}+2\mathbf{b})$
17 $\frac{1}{2}(\mathbf{b}+\mathbf{c})$, $-\mathbf{b}+\frac{1}{2}\mathbf{c}$, $\frac{1}{2}\mathbf{b}-\mathbf{c}$
18 $\sqrt{20}$
19 $\mathbf{c}=\mathbf{a}+2\mathbf{b}$, 2 : 5
20 $\frac{6}{7}$, $\frac{3}{7}$, $\frac{2}{7}$

Exercise 3

1 12
2 (a) -3, (b) $\frac{4}{3}$, (c) 10, $-\frac{2}{5}$
3 $\pm\sqrt{2}$
4 $\frac{26}{17}$
5 $\sqrt{2}$
7 (a) (i) π, (ii) $2\pi/3$, (iii) $2\pi/3$ (b) $\lambda=-14$, $\mu=12$
9 $6\mathbf{i}+2\mathbf{j}$, $6\mathbf{j}$
10 9; $\pi/2$, arc cos $(1/\sqrt{5})$, arc cos $(2/\sqrt{5})$
11 0, $\sqrt{\frac{2}{3}}$, $\sqrt{\frac{1}{3}}$
13 $AP^2=(1-k)^2(\mathbf{a}^2+\mathbf{b}^2-2\mathbf{a}.\mathbf{b})$, $OP^2=k^2\mathbf{a}^2+(1-k)^2\mathbf{b}^2+2k(1-k)\mathbf{a}.\mathbf{b}$
14 $\mathbf{a}=-\mathbf{i}+3\mathbf{j}$, $\mathbf{b}=13\mathbf{i}-11\mathbf{j}$
$\mathbf{c}=-10\mathbf{s}+10\sqrt{2}\mathbf{t}$
15 $(\mathbf{p}+\mathbf{q}+4\mathbf{r})/\sqrt{18}$

Exercise 4

1 $\frac{1}{4}(\mathbf{p}+3\mathbf{q})$
2 $\frac{22}{3}\mathbf{i}+7\mathbf{j}$
4 $\frac{1}{11}(\mathbf{a}+6\mathbf{b}+4\mathbf{c})$
5 $\mathbf{a}+\mathbf{b}+\mathbf{c}$
6 $6\mathbf{a}-5\mathbf{b}$
7 $3\mathbf{a}+3\mathbf{b}$; (i) $\mathbf{r}=3(\mathbf{a}+\mathbf{b})+\lambda\mathbf{c}$ where $\mathbf{c}.\mathbf{a}=\mathbf{c}.\mathbf{b}=0$, (ii) $\mathbf{r}=3(\mathbf{a}+\mathbf{b})+\lambda\mathbf{a}+\mu\mathbf{b}$
8 $\mathbf{r}=\lambda(\mathbf{i}-\mathbf{j}+3\mathbf{k})+(1-\lambda)(\mathbf{i}+2\mathbf{j}+2\mathbf{k})$, $\frac{1}{3}(3\mathbf{i}+3\mathbf{j}+7\mathbf{k})$
9 $\mathbf{n}=\frac{1}{3}(2, 2, 1)$, $p=3$
10 $\mathbf{i}-\mathbf{j}+2\mathbf{k}$,
(a) $\mathbf{r}=\mathbf{i}-\mathbf{j}+2\mathbf{k}+\lambda(3\mathbf{i}+8\mathbf{j}-2\mathbf{k})$,
(b) $[\mathbf{r}-(\mathbf{i}-\mathbf{j}+2\mathbf{k})].(4\mathbf{i}-\mathbf{j}+2\mathbf{k})=0$,
(c) $\mathbf{r}=(\mathbf{i}-\mathbf{j}+2\mathbf{k})+\lambda(\mathbf{i}-2\mathbf{k})$

11 $[\mathbf{r} - (\mathbf{i} - \mathbf{j} + 2\mathbf{k})]$
$.(\mathbf{i} + 2\mathbf{j} + 5\mathbf{k}) = 0$, $1/\sqrt{30}$

12 $4\mathbf{i} + 7\mathbf{j} - 5\mathbf{k}$, (a) $2\mathbf{i} + \mathbf{j} - \mathbf{k}$,
(b) $2\sqrt{14}$

13 $\mathbf{r} = \frac{1}{11}(27\mathbf{i} + 19\mathbf{j} + 14\mathbf{k})$
$+ \lambda(\mathbf{i} - 3\mathbf{j} - \mathbf{k})$,
$\frac{1}{11}(26\mathbf{i} + 22\mathbf{j} + 15\mathbf{k})$

14 $\frac{1}{4}(\mathbf{a} + \mathbf{b} + \mathbf{c} + \mathbf{d})$

15 (1, 1, 2), arc cos $(\frac{1}{3})$

16 (a) 10, arc cos $(2/\sqrt{5})$, (b) $4\sqrt{5}$,
arc cos $(1/\sqrt{5})$ with Ox

17 $(12\mathbf{i} + 10\mathbf{j})$ km

18 (a) $(110\mathbf{i} - 5\mathbf{j})$ km,
(b) $(240\mathbf{i} - 100\mathbf{j})\ \text{km h}^{-1}$,
(c) $(30\mathbf{i} - 380\mathbf{j})\ \text{km h}^{-1}$

19 $(2\mathbf{i} - \mathbf{j})\ \text{m s}^{-1}$, $(5\mathbf{i} + 2\mathbf{j})$ m

20 (a) $(-3\mathbf{i} + 4\mathbf{j})\ \text{m s}^{-1}$,
(b) $(-\mathbf{i} + 15\mathbf{j}) + t(6\mathbf{i} - 2\mathbf{j})$;
$\sqrt{193.6}$ km, $\frac{9}{10}\sqrt{13}$ km

21 (i) $(2t\mathbf{i} + 6t\mathbf{j})$ m,
$[(4t + \frac{1}{2}t^2)\mathbf{i} + (-8t + \frac{1}{2}t^2)\mathbf{j}]$ m,
(ii) 5 s, (iii) 6 s

22 $(-2\mathbf{i} + \mathbf{j})\ \text{m s}^{-1}$, $4/\sqrt{5}$ m

23 (ii) $(3\mathbf{i} + 4\mathbf{j})\ \text{m s}^{-1}$

24 $(12\mathbf{i} + 6\mathbf{j})$ N

25 $(6\mathbf{i} - 3\mathbf{j})$ N

26 (a) $\sqrt{122}$ N, (b) 11

27 $-\frac{9}{4}\mathbf{i}$ m

28 2, 8 N

29 $\frac{1}{4}(29\mathbf{i} - 11\mathbf{j} + 54\mathbf{k})$ m

30 $p = 9$, $q = 12$

31 $(27\mathbf{i} + 80\mathbf{j})$ N

32 $\sqrt{229}$ N

Exercise 5

1 (a) $8t\ \text{m s}^{-1}$, $8\ \text{m s}^{-1}$, (c) $21\mathbf{i}\ \text{m s}^{-2}$

2 (a) $(9\mathbf{i} + 6\mathbf{j})$ m, $(9\mathbf{i} - \mathbf{j})$ m, 7 m
(b) $(2\mathbf{i} + 2\mathbf{j})\ \text{m s}^{-1}$,
(c) $(-\mathbf{i} + 2\mathbf{j})\ \text{m s}^{-1}$

3 (a) $9\mathbf{i}\ \text{m s}^{-1}$, $(2\mathbf{i} + \mathbf{j})\ \text{m s}^{-1}$,
(b) $(7\mathbf{i} - \mathbf{j})\ \text{m s}^{-1}$, (c) $3\mathbf{i}\ \text{m s}^{-2}$,
$(3\mathbf{i} - 2\mathbf{j})\ \text{m s}^{-2}$

4 $x^2 + y^2 = 4a^2$, $x^2 = 4ay$
(a) 0,
(b) $2\sqrt{2}a\omega^2$, (c) $(2n + 1)\pi/\omega$
(d) $2n\pi/\omega$

6 $s : a(\cos \omega t\,\mathbf{i} + \sin \omega t\,\mathbf{j})$,
$t : a(-\sin \omega t\,\mathbf{i} + \cos \omega t\,\mathbf{j})$,
$a\omega(s - c)\mathbf{i} - a\omega(s + c)\mathbf{j}$

7 $\tan \theta = \frac{1}{2}t$

8 $\mathbf{r} = \mathbf{V}t + \frac{1}{2}\mathbf{f}t^2$

9 (a) $(10\mathbf{i} - 10\mathbf{j} + 20\mathbf{k})\ \text{m s}^{-1}$,
$(12\mathbf{i} - 6\mathbf{j} + 16\mathbf{k})\ \text{m s}^{-2}$
(b) $[(3\mathbf{i} + \mathbf{j} + 4\mathbf{k}) + (-2e^{-t}\mathbf{i}$
$+ 5 \sin t\,\mathbf{j} - 3 \cos t\,\mathbf{k})]\ \text{m s}^{-1}$,
$[(-\mathbf{i} + 8\mathbf{j} + \mathbf{k}) + (3\mathbf{i} + \mathbf{j} + 4\mathbf{k})t$
$+ (2e^{-t}\mathbf{i} - 5 \cos t\,\mathbf{j} - 3 \sin t\,\mathbf{k})]$ m

11 $\dfrac{d\mathbf{b}}{dt} = -\mathbf{a}\dfrac{d\theta}{dt}$; $-r\dot{\theta}^2$, $r\ddot{\theta}$

Exercise 6

1 (a) 24, (b) $6\mathbf{i} + 18\mathbf{j} + 6\mathbf{k}$

3 $\sqrt{(101/17)}$

4 $p = q = s = -1$, $\cos \theta = -2\sqrt{2}/3$

6 $\mathbf{v} = a\omega(\mathbf{i} \cos \omega t - \mathbf{j} \sin \omega t + \mathbf{k})$,
$\mathbf{f} = -a\omega^2(\mathbf{i} \sin \omega t + \mathbf{j} \cos \omega t)$
$t = n\pi/\omega$

8 $\mathbf{r} = \lambda\mathbf{a} + (\mathbf{a} \times \mathbf{b})/\mathbf{a}^2$

9 $\frac{4}{11}(-34\mathbf{i} + 51\mathbf{j})$ N m

10 $(11\mathbf{i} + 8\mathbf{j} - 3\mathbf{k})$ N m

11 $c = 3$, $\mathbf{r} = \mathbf{i} + 3\mathbf{j} + \lambda(3\mathbf{i} + 2\mathbf{j} + 4\mathbf{k})$

12 $\pm\dfrac{1}{\sqrt{50}}(3\mathbf{i} + 5\mathbf{j} + 4\mathbf{k})$,
$3x + 5y + 4z = 30$, $3\sqrt{2}$.

13 (a) the line AB,
(b) the plane through A, B, C,
(c) the plane normal to AB through the mid-point of AB,
(d) the line through the centre of the circle through ABC and perpendicular to the triangle ABC,
(e) the line through the origin normal to triangle ABC. (b) and (d).

INDEX